# Springer
# Proceedings in Physics    53

# Springer Proceedings in Physics

Managing Editor: H.K.V. Lotsch

30 *Short-Wavelength Lasers and Their Applications* Editor: C. Yamanaka

31 *Quantum String Theory*
Editors: N. Kawamoto and T. Kugo

32 *Universalities in Condensed Matter*
Editors: R. Jullien, L. Peliti, R. Rammal, and N. Boccara

33 *Computer Simulation Studies in Condensed Matter Physics: Recent Developments*
Editors: D. P. Landau, K. K. Mon, and H.-B. Schüttler

34 *Amorphous and Crystalline Silicon Carbide and Related Materials*
Editors: G. L. Harris and C. Y.-W. Yang

35 *Polycrystalline Semiconductors: Grain Boundaries and Interfaces*
Editors: H. J. Möller, H. P. Strunk, and J. H. Werner

36 *Nonlinear Optics of Organics and Semiconductors*
Editor: T. Kobayashi

37 *Dynamics of Disordered Materials*
Editors: D. Richter, A. J. Dianoux, W. Petry, and J. Teixeira

38 *Electroluminescence*
Editors: S. Shionoya and H. Kobayashi

39 *Disorder and Nonlinearity*
Editors: A. R. Bishop, D. K. Campbell, and S. Pnevmatikos

40 *Static and Dynamic Properties of Liquids*
Editors: M. Davidović and A. K. Soper

41 *Quantum Optics V*
Editors: J. D. Harvey and D. F. Walls

42 *Molecular Basis of Polymer Networks*
Editors: A. Baumgärtner and C. E. Picot

43 *Amorphous and Crystalline Silicon Carbide II: Recent Developments*
Editors: M. M. Rahman, C. Y.-W. Yang, and G. L. Harris

44 *Optical Fiber Sensors*
Editors: H. J. Arditty, J. P. Dakin, and R. Th. Kersten

45 *Computer Simulation Studies in Condensed Matter Physics II: New Directions*
Editors: D. P. Landau, K. K. Mon, and H.-B. Schüttler

46 *Cellular Automata and Modeling of Complex Physical Systems*
Editors: P. Manneville, N. Boccara, G. Y. Vichniac, and R. Bidaux

47 *Number Theory and Physics*
Editors: J.-M. Luck, P. Moussa, and M. Waldschmidt

48 *Many-Atom Interactions in Solids*
Editors: R. M. Nieminen, M. J. Puska, and M. J. Manninen

49 *Ultrafast Phenomena in Spectroscopy*
Editors: E. Klose and B. Wilhelmi

50 *Magnetic Properties of Low-Dimensional Systems II: New Developments*
Editors: L. M. Falicov, F. Mejía-Lira, and J. L. Morán-López

51 *The Physics and Chemistry of Organic Superconductors*
Editor: G. Saito and S. Kagoshima

52 *Dynamics and Patterns in Complex Fluids: New Aspects of the Physics-Chemistry Interface*
Editors: A. Onuki and K. Kawasaki

53 *Computer Simulation Studies in Condensed Matter Physics III*
Editors: D. P. Landau, K. K. Mon, and H.-B. Schüttler

Volumes 1–29 are listed on the back inside cover

# Computer Simulation Studies in Condensed Matter Physics III

Proceedings of the Third Workshop
Athens, GA, USA, February 12–16, 1990

Editors: D. P. Landau, K. K. Mon, and H.-B. Schüttler

With 101 Figures

Springer-Verlag
Berlin Heidelberg New York
London Paris Tokyo
Hong Kong Barcelona
Budapest

Professor David P. Landau, Ph.D.
Professor K. K. Mon, Ph.D.
Professor Heinz-Bernd Schüttler, Ph.D.
Center for Simulation Physics,
The University of Georgia,
Athens, GA 30602, USA

ISBN 3-540-53607-8 Springer-Verlag Berlin Heidelberg New York
ISBN 0-387-53607-8 Springer-Verlag New York Berlin Heidelberg

This work is subject to copyright. All rights are reserved, whether the whole or part of the material is concerned, specifically the rights of translation, reprinting, reuse of illustrations, recitation, broadcasting, reproduction on microfilms or in other ways, and storage in data banks. Duplication of this publication or parts thereof is only permitted under the provisions of the German Copyright Law of September 9, 1965, in its current version, and a copyright fee must always be paid. Violations fall under the prosecution act of the German Copyright Law.

© Springer-Verlag Berlin Heidelberg 1991
Printed in Germany

The use of registered names, trademarks, etc. in this publication does not imply, even in the absence of a specific statement, that such names are exempt from the relevant protective laws and regulations and therefore free for general use.

54/3140 – 543210 – Printed on acid-free paper

# Preface

The contribution of computer simulation studies to our understanding of the properties of a wide range of condensed matter systems is now well established. The Center for Simulational Physics of the University of Georgia has been hosting a series of annual workshops with the intent of bringing together experienced practitioners in the field, as well as relative newcomers, to provide a forum for the exchange of ideas and recent results. This year's workshop, the third in the series, was held February 12–16, 1990. These proceedings are a record of the workshop and are published with the goal of timely dissemination of the papers to a wider audience.

The proceedings are divided into four parts. The first contains invited papers dealing with simulational studies of classical systems and also includes an introduction to some new simulation techniques. A separate section is devoted to invited papers on quantum systems, including new results for strongly correlated electron and quantum spin models believed to be important for the description of high-$T_c$ superconductors. The third part consists of a single invited paper, which presents a comprehensive treatment of issues associated with high performance computing, including differences in architectures and a discussion of access strategies. The contributed papers constitute the final part.

We hope that readers will benefit from papers in their own specialty as well as profit from exposure to new methods and ideas. We have already learned from preceding years that fruitful collaborations and new research projects have resulted from these gatherings at the University of Georgia, and we hope that the proceedings may further foster collaborative and possibly interdisciplinary research.

This workshop was made possible through the generous support of the Center for Simulational Physics and the Vice President for Research at the University of Georgia.

Athens, GA
April 1990

*D.P. Landau*
*K.K. Mon*
*H.-B. Schüttler*

# Contents

Computer Simulation Studies in Condensed Matter Physics:
An Introduction
By D.P. Landau, K.K. Mon, and H.-B. Schüttler ................ 1

## Part I  Classical Systems

Simulation of Order–Disorder Phenomena and Diffusion
in Metallic Alloys
By K. Binder (With 13 Figures) ............................ 4

Monte Carlo Analysis of the Ising Model and CAM
By N. Ito and M. Suzuki (With 2 Figures) .................... 16

Histogram Techniques for Studying Phase Transitions
By A.M. Ferrenberg (With 3 Figures) ....................... 30

Classification of Cellular Automata
By R.W. Gerling ......................................... 43

Simulation Studies of Classical and Non-classical Nucleation
By W. Klein (With 9 Figures) .............................. 50

Molecular Dynamics of Slow Viscous Flows
By J.R. Banavar, J. Koplik, and J.F. Willemsen (With 11 Figures) ..... 65

Computer Simulations for Polymer Dynamics
By K. Kremer, G.S. Grest, and B. Dünweg (With 7 Figures) ......... 85

Computer Simulation Studies of Phase Transitions in Two-Dimensional
Systems of Molecules with Internal Degrees of Freedom
By O.G. Mouritsen, D.P. Fraser, J. Hjort Ipsen, K. Jørgensen,
and M.J. Zuckermann (With 11 Figures) ...................... 99

## Part II  Quantum Systems

Numerical Evaluation of Candidate Wavefunctions
for High-Temperature Superconductivity
By R. Joynt ............................................. 116

Two-Dimensional Quantum Antiferromagnet at Low Temperatures
By E. Manousakis (With 10 Figures) .......................... 123

Binding of Holes in the Hubbard Model
By A. Moreo (With 4 Figures) ............................... 138

The Average Spectrum Method for the Analytic Continuation
of Imaginary-Time Data
By S.R. White (With 6 Figures) .............................. 145

## Part III  High Performance Computing

High Performance Computing in Academia: A Perspective
By W.B. McRae ............................................ 156

## Part IV  Contributed Papers

Finite-Size Scaling Study of the Simple Cubic Three-State Potts Glass
By M. Scheucher, J.D. Reger, K. Binder, and A.P. Young
(With 4 Figures) ........................................... 172

Numerical Transfer Matrix Studies of Ising Models
By M.A. Novotny (With 2 Figures) ............................ 177

A Computer Simulation of Polymers with Gaussian Couplings:
Modeling Protein Folding?
By H.-O. Carmesin and D.P. Landau (With 3 Figures) ............. 183

Numerical Studies of Absorption of Water in Polymers
By J.L. Vallés, J.W. Halley, and B. Johnson (With 6 Figures) ........ 187

Almost Markov Processes
By D. Bouzida, S. Kumar, and R.H. Swendsen .................. 193

Vectorization of Diffusion of Lattice Gases
Without Double Occupancy of Sites
By O. Paetzold (With 1 Figure) .............................. 197

Ising Machine, m-TIS
By N. Ito, M. Taiji, M. Suzuki, R. Ishibashi, K. Kobayashi, N. Tsuruoka,
and K. Mitsubo ........................................... 201

Monte Carlo Analysis of Finite-Size Effects
in the Three-Dimensional Three-State Potts Model
By O.F. de Alcantara Bonfim (With 4 Figures) ................... 203

Simulation Studies of Oxygen Ordering in $YBa_2Cu_3O_{7-\delta}$
and Related Systems
By Zhi-Xiong Cai and S.D. Mahanti (With 5 Figures) ............. 210

**Index of Contributors** ..................................... 215

# Computer Simulation Studies in Condensed Matter Physics: An Introduction

*D.P. Landau, K.K. Mon, and H.-B. Schüttler*

Center for Simulational Physics, The University of Georgia,
Athens, GA 30602, USA

This year's workshop includes contributions which deal with several different aspects of the rapidly developing area of computer simulations in condensed matter physics. The invited talks are reported in rather long papers which have some pedagogical content, and at the end of these proceedings we present a few shorter, contributed papers which present very recent research results.

The first portion of this volume contains invited papers which deal with classical systems. In the first of these, Binder reviews the application of the Monte Carlo method to the study of lattice models in metallurgy. The static behavior of bcc and fcc binary alloy models is determined and the application to real systems like Cu–Au and Fe–Al is discussed. Studies of diffusion in alloys and the initial stages of spinodal decomposition are also treated.

Ito and Suzuki develop a cluster–effective–field approximation which is used to study phase transitions in Ising ferromagnets on square and simple cubic lattices. Using Monte Carlo simulations they calculate the coherent anomaly method coefficients and estimate the critical temperature and susceptibility with good accuracy.

The application of histogram methods to the study of critical phenomena is presented by Ferrenberg. After describing the theoretical foundation of the technique, he points out some computational difficulties which may be encountered and shows how they may be circumvented. Results obtained for three models are briefly reviewed: a 2d dimer model, the 3d Heisenberg model, and the 3d Ising model.

Gerling presents the results of a study of cellular automata on lattices with dimensionality between 1 and 3. Details of the algorithm are given as well as results for classification and damage spreading for these models with different rules. For coordination number $q<6$ all rules were used, but for $q=6$ only a random sampling of rules was possible.

Klein presents a review of simulation studies of nucleation in Ising models with several ranges of interactions. The results indicate that the classical theory of nucleation correctly describes the properties of systems with short range coupling. In contrast, the data for systems with long range interactions undergoing deep quenches is not consistent with predictions from classical theory.

The paper by Banavar, Koplik, and Willemsen presents a molecular dynamics (MD) study of slow viscous flows. The emphasis is on using the MD technique as a bridge between the microscopic and macroscopic levels of description. Model systems of several thousand molecules are shown to exhibit reasonable continuum behavior.

Kremer, Grest and Dünweg discuss recent work on the dynamics of polymeric systems. Monte Carlo and MD methods were used in 2d and 3d systems. Comparison with experiment shows that the time and length scales for the onset of reptation can be correctly predicted from computer simulations for a variety of polymeric liquids.

Models of 2d systems of molecules with internal degrees of freedom are discussed by Mouritsen, Fraser, Ipsen, Jørgensen, and Zuckermann. These are used to describe phase transitions in lipid mono– or bi–layer systems. The interplay between ordering processes governed by different degrees of freedom has been studied by large scale simulations.

The second part of this volume contains several papers on simulations of quantum systems. The paper by Joynt gives a review of recent variational Monte Carlo studies of 2d Hubbard and t–J models using Gutzwiller–type wavefunctions. Of particular interest here are results for the antiferromagnetic spin fluctuations in the almost localized Fermi liquid ground state and the competition between normal Fermi liquid, magnetic, and various possible superconducting ground states which have been investigated by such variational techniques.

Manousakis discusses recent variational and quantum Monte Carlo simulation studies of the 2d s=1/2 Heisenberg antiferromagnet and its relationship to the quantum non–linear $\sigma$–model. New results for the dynamics of a dopant induced hole carrier in the 2d t–J model have been obtained on the basis of a novel variational wavefunction which includes spin backflow effects.

The paper by Moreo describes a novel application of the determinental Fermion Monte Carlo method to measure ground–state energy differences and hole–hole binding energies in the 2d Hubbard model near half filling. The basic idea here is to simulate in a grand–canonical ensemble with a purely imaginary chemical potential. There is then no minus sign problem and the desired energy difference can be directly extracted from simulated intensive quantities.

Significant progress has recently been made in extracting real–frequency spectral information from imaginary–time dynamical correlation function data which can be obtained by quantum Monte Carlo simulations. The paper by White describes one of these recent developments and discusses results obtained with this method for the single–particle density of states and for the dynamics of superconducting pair susceptibilities in the two-dimensional Hubbard model.

In a single invited paper comprising the third section of these proceedings, McRae reviews some of the history of research use of supercomputers and surveys current architectures, compilers, and methods of performance evaluation. He summarizes products on the market today and describes the consideration of access strategies carried out by a University of Georgia task force.

In the last part we present the texts of a series of contributed papers. Scheucher, Reger, Binder, and Young present a finite size scaling study of the simple cubic three–state Potts–glass. From extensive Monte Carlo simulations these authors conclude that the Potts–glass is different from the 3d Ising–glass. Novotny considers a numerical transfer matrix study of the Ising model in high dimensions. Results for the critical temperature and thermal critical exponent are presented for dimensions up to and including seven. Carmesin and Landau describe a simulation study of polymers with Gaussian couplings. Data show that a random Gaussian heteropolymer folds into specific preferred conformations at some temperature and exhibits hysteresis while unfolding. Vallés, Halley, and Johnson report on two computer simulation studies of water penetration into polymers. An MD study is used to examine the bonding of a water molecule to a polymer of industrial interest, and a Monte Carlo simulation is used to study diffusion with trapping on a percolation lattice. Bouzida, Kumar, and Swendsen present a new Dynamically Optimized Monte Carlo method for making simulations of inhomogeneous, anisotropic systems more efficient. Paetzold presents a vectorizable algorithm for the Monte Carlo simulation of diffusion in lattice gases. The method is used to study a 3d percolation lattice. Ito describes a special purpose computer for the Monte Carlo simulation of simple Ising models. De Alcantara Bonfim uses the histogram method (described by Ferrenberg earlier in these proceedings) to study the three state Potts models in three dimensions. Cai and Mahanti have carried out simulations on lattice gas models for oxygen ordering in the $YBa_2Cu_3O_{7-\delta}$ high $T_c$ superconductor to study the effects of thermal quenching and substitutional impurities at the Cu chain sites.

Part I

**Classical Systems**

# Simulation of Order–Disorder Phenomena and Diffusion in Metallic Alloys

*K. Binder*

Materialwissenschaftliches Forschungszentrum (MWFZ) and
Institut für Physik, Johannes-Gutenberg-Universität Mainz,
Staudinger Weg 7, W-6500 Mainz, Fed. Rep. of Germany

Abstract: The application of the Monte Carlo method to lattice–statistics problems in metallurgy is reviewed. Examples are given for the prediction of phase diagrams from simple model assumptions for effective interatomic potentials and for the calculation of parameters describing long– and short–range order, ordering energy, etc., both for face–centered cubic (fcc) and body–centered cubic (bcc) lattices. Applications to real systems such as Cu–Au and Fe–Al alloys are discussed.

Then studies of diffusion in alloys based on a simple vacancy model are presented, and the relation of the interdiffusion constant to the jump rates $\Gamma_A$, $\Gamma_B$ of the two species of atoms is discussed for the non–interacting case. Also the modeling of the initial stages of spinodal decomposition is discussed briefly.

In summary, it is shown that the statics and dynamics of ordering phenomena pose many interesting and nontrivial problems in statistical mechanics, and that Monte Carlo simulations yield a lot of insight into such problems, and also contribute to a better understanding of experiments and real materials.

## 1. Introduction

Understanding order–disorder phase transitions in metallic alloys has been a problem of interest for both metallurgy and statistical physics for decades [1–25]. E.g., alloys such as $\beta$–brass have served as model systems to experimentally establish [25] key concepts of the theory of critical phenomena (non–mean–field critical exponents, scaling, universality); other alloys, such as the Cu–Au–system, serve as a testing ground for methods dealing with the statistical mechanics of first–order phase transitions and phase diagram calculations [2,3,5–7, 9,11,14–17,19,23]. For both types of problems, it is of interest to establish a quantitative link between microscopic models of the effective interactions between the ions in the alloy and macroscopic material properties. Monte Carlo computer simulation methods [26–28] can make a useful contribution to such efforts.

Also dynamic phenomena associated with ordering of atoms in alloys are of great physical interest: often the time scale of such processes, which is determined by the vacancy mechanism of diffusion in solids [29,30], is so slow that the initial stages of interdiffusion, formation of ordered domains, or unmixing after quenching experiments can be followed by experiments in real time. Consequently, metallic alloys have played a key role in elucidating mechanisms of phase transformations such as nucleation and spinodal decomposition [31–33]. Monte Carlo simulations, in turn, have contributed significantly to the understanding of such processes [35] and of diffusion phenomena in general [36–42].

This contribution illustrates this interplay between Monte Carlo simulation and both theory and experiment with some characteristic examples, taken from the research work of the author's group. Thus no exhaustive review is given, but rather the spirit of the approach to these problems on the statics and dynamics of ordering phenomena in alloys is demonstrated.

## 2. Ordering of Face–Centered Cubic Alloys and Application to the Cu–Au–System

Fig. 1 shows a partial experimental [43] phase diagram of the copper–gold system in the temperature – concentration plane. The curves indicate the existence regions of the four ordered phases $Cu_3Au$, $CuAu_3$, CuAuI and CuAuII. These phases are separated from each other (and from the disordered phase occurring at higher temperatures) by two–phase coexistence regions, since we are dealing with first–order phase transitions throughout [32]. Note that the

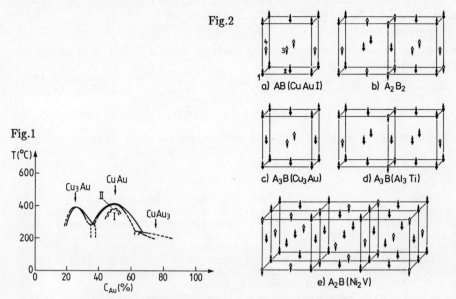

Fig.1:   Partial phase diagram of Cu–Au alloys. From Ref. [43]

Fig.2:   Ordered structures of substitutional binary alloys on the face–centered cubic (fcc) lattice. From Ref. [11]

phase boundaries cannot be measured at temperatures lower than drawn, since there diffusion is too slow to allow thermal equilibrium to be established during physically accessible time scales.

The ordered structures CuAuI (or symbolically AB in the general binary alloy A–B) and Cu$_3$Au (A$_3$B) are shown in Fig. 2, together with some other ordered superstructures occurring on fcc alloys (the CuAuII structure is a long period modulated version of the simple CuAuI structure, Fig. 2a), and is not shown). A–atoms (black dots) can also be represented in an Ising spin model representation [11,23] as "spin up", B–atoms (open circles) as "spin down". Decomposing one cube of the fcc lattice into four interpenetrating sublattices 1, 2, 3, 4 as shown in Fig. 2a, the AB structure means an ordering where two sublattices are "spin up" and two sublattices are "spin down" (↑↑↓↓), while the A$_3$B ordering means that three sublattices are "spin up", one is "spin down" (↑↑↑↓). If the simple cubic sublattices themselves are split into different sublattices in various ways, structures like A$_2$B$_2$ and A$_3$B of Al$_3$Ti–type can be realized, which have larger unit cells {Figs. 2b), d), e)}.

Now the question which we want to discuss is: to what extent can ordering phenomena such as shown in Figs. 1,2 be explained in terms of simple lattice models, where we treat the lattice as both perfect (neglecting lattice defects such as vacancies, dislocations, grain boundaries etc.) and rigid (neglecting lattice vibrations), and consider only the configurational problem where the state of the alloy is described in terms of local occupation variables $c_i^A$, $c_i^B$ ($= 1 - c_i^A$). Here $c_i^A = 1$ if site i is occupied by an A–atom and $c_i^A = 0$ otherwise. Assuming for simplicity pairwise interactions only, the Hamiltonian $\mathcal{H}$ of the system is

$$\mathcal{H} = \mathcal{H}_0 + \sum_{<i,j>} \left[ c_i^A c_j^A v_{AA}(\vec{r}_i - \vec{r}_j) + 2 c_i^A c_j^B v_{AB}(\vec{r}_i - \vec{r}_j) + c_i^B c_j^B v_{BB}(\vec{r}_i - \vec{r}_j) \right], \quad (1)$$

$\mathcal{H}_0$ being some background term, and $v_{AA}$, $v_{AB}$, $v_{BB}$ being potential energies between AA, AB and BB pairs (the symbol <i,j> means that over such pairs is summed once). As is well known, Eq. (1) is isomorphic to an Ising problem in a magnetic field H: via the transformation $S_i \equiv 1 - 2c_i^B = \pm 1$ we can reduce $\mathcal{H} - \Delta\mu N_B \equiv \mathcal{H} - (\mu_B - \mu_A) N_B$ where $N_B = \sum_i c_i^B = <c^B>N$, N being the number of lattice sites, to the Ising model

$$\mathcal{H}_{\text{Ising}} = -\sum_{<i,j>} J(\vec{r}_i - \vec{r}_j) S_i S_j - H \sum_i S_i,  \qquad (2)$$

where constant terms have been omitted and the "exchange constant" is related to the effective ordering potential $\Delta v$ as

$$2J(\vec{r}_i - \vec{r}_j) = \Delta v(\vec{r}_i - \vec{r}_j) \equiv$$
$$\equiv v_{AB}(\vec{r}_i - \vec{r}_j) - [v_{AA}(\vec{r}_i - \vec{r}_j) + v_{BB}(\vec{r}_i - \vec{r}_j)]/2, \qquad (3)$$

and the "magnetic field" is related to the difference in chemical potentials $\Delta\mu$ as

$$2H - \sum_{j(\neq i)} \left[ v_{AA}(\vec{r}_1 - \vec{r}_j) - v_{BB}(\vec{r}_1 - \vec{r}_j) \right] - \Delta\mu \qquad (4)$$

It is not clear at all a priori, whether such a simple model is a reasonable starting point for discussing metallic alloys such as CuAu [21,23], or whether one should better start from electronic structure theory [44,45]. Now the comparison between any analytical theory for this problem and experiment is hampered by the fact that analytical theories require drastic approximations, the accuracy of which is completely unknown! Even for the simplest case, where Eqs. (2)–(4) are used and the range of the interaction in Eq. (3) is restricted to nearest neighbors (nn), the problem is very difficult: the resulting phase diagram of this very simple model strongly depends on the approximations (Fig. 3).

At this level of statistical approximation, it is obviously premature to discuss whether the model Eq. (2)–(4), which leads to phase diagrams which are necessarily symmetrical around the line $c_B = 1/2$ due to the spin reversal symmetry of the Ising model for $H = 0$, should be made

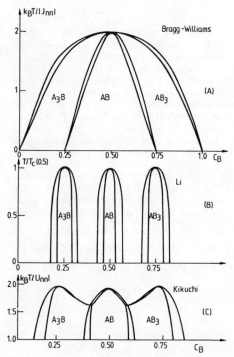

Fig.3: Temperature–concentration phase diagram of a binary alloy A–B at the fcc lattice with nearest–neighbor interaction $J_{nn} < 0$ according to the Bragg–Williams approximation (A) [2], the quasi–chemical approximation (B) [3] and the cluster variation method [3] in the tetrahedron approximation [5,6] (C). From Ref. [7].

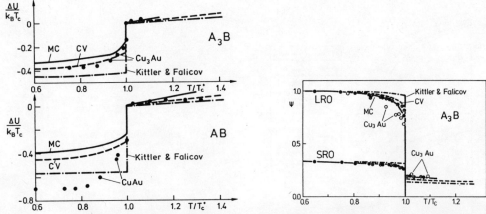

Fig.4: Ordering energy ΔU (normalized to zero in the disordered state) plotted vs. temperature for the $A_3B$ alloy and the AB alloy (left) and long range order parameter (LRO) and short range order parameter $-\alpha_1$ (SRO) plotted vs. temperature (right). From Ref. [9].

more realistic by introducing three– and four–body interactions [7], or by introducing further neighbor interactions $J_{nnn}$, $J_3$ etc. with various $R \equiv J_{nnn}/J_{nn}$, $R_3 = J_3/J_{nn}$ etc. depending on the average concentration $c_B$ [23], or both. The latter suggestion [23] of effective smoothly concentration–dependent interactions is motivated by the experimental finding [46] that $\Delta v(r)$ typically has a Friedel form,

$$\Delta v(r) = A \, cos(2|\vec{k}_F|r + \Phi)/r^3, \qquad (5)$$

where A, $\Phi$ are constants and $\vec{k}_F$ is the Fermi wave vector. Now both $|\vec{k}_F|$ and the lattice spacing differ somewhat when one goes from Cu to Au, of course, and particularly for larger distances, where the argument of the *cos* function far exceeds $\pi$, a significant concentration dependence of $\Delta v(r)$ is expected.

The fact that a very accurate statistical–mechanical treatment of the model is required is also very clearly recognized when we attempt a comparison of theoretical predictions for the ordering energy $\Delta U = 2 \sum_{<i,j>} <c_i^A c_j^B> \Delta v(\vec{r}_i - \vec{r}_j)$ and parameters $\psi$, $\alpha_l$ characterizing long range order ($\psi$) and short range order (SRO) in the l'th neighbor shell ($\alpha_l$) to corresponding experimental data (Fig. 4). Here the Cowley [47] SRO parameters $\alpha_l$ are defined as

$$\alpha(\vec{r}_i - \vec{r}_j) = \frac{<c_i^B c_j^B> - c_B^2}{c_B(1 - c_B)} = \frac{<S_i S_j> - m^2}{1 - m^2} \qquad (6)$$

where $m = <S_i>$. It is seen that the treatment based on electronic structure theory [45] deviates from the data very strongly, and thus complicated theories of this type do not constitute any practical advantage at all over the simple Ising–type lattice models considered here. On the other hand, in most cases the deviation between the calculations based on the cluster variation (CV) method [5,6] and the Monte Carlo (MC) method [9], which refer to exactly the same nearest neighbor model, are of the same order as the deviation between the MC results (which are believed to be closest to the exact answer) and the experimental data [48–51]. This fact reiterates the claim [9,11,23] that not much can be learnt if an inaccurate approximation of a simplified model is compared to experiment: it is then necessarily unclear whether discrepancies should be attributed to inaccuracies of the approximation, simplifications of the model or both! Also, an agreement between data and inaccurate calculations may be purely accidental, as the example of the ordering energy of $Cu_3Au$ for $T < T_c$ [48] shows (Fig. 4). Thus clearly a rather accurate statistical mechanics is required, if one wants to improve the

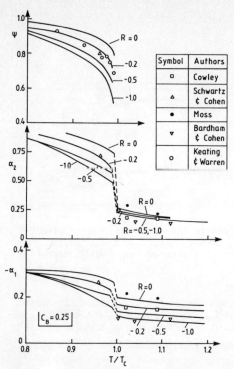

Fig.5: Temperature dependence of the long range order parameter $\psi$ of the $Cu_3Au$ system and of the Cowley short range order parameters $\alpha_1$, $\alpha_2$ compared to model calculations for various values R of the ratio between the ordering potential $\Delta v$ between next nearest and nearest neighbors. Curves are Monte Carlo results [23], points show experimental data from Refs. [49–53] as indicated. From Ref. [23].

model by studying the effects of three– and four–body forces or smaller interactions between more distant neighbors. The Monte Carlo method so far is the <u>only</u> approach which yields the necessary accuracy for the present problem. As an example, Fig. 5 shows the temperature dependence of order parameters of $Cu_3Au$ [49–53] compared to MC results, with various choices of the ratio R between next–nearest and nearest neighbor interactions. It is seen that for $T < T_c$ the data <u>for all quantities simultaneously</u> are consistent with $R \approx -0.2$. This is reasonable since for this value also the ratio of the transition temperature at the triple point to the $Cu_3Au$ transition temperature is about 0.89 [12] while experimentally it is 0.84, see Fig. 1, which is not far off (the experimental value would be reproduced for $R \approx -0.17$ [54]). Unfortunately, the short range order data for $T > T_c$ do not allow any conclusion: the scatter of the data taken from different authors is larger than the variation of $\alpha_1$, $\alpha_2$ with R from $R = 0$ to $R = -1$! Although this comparison is somewhat unfair – some of the early work could not be corrected for the effect of lattice vibrations, scattering from dislocations etc. and hence is systematically in error – it is fair to say that what is most urgently needed to make further progress in the understanding of Cu–Au alloy ordering behavior are more complete and reliable experimental data.

## 3. Model Calculations of Phase Diagrams of Magnetic Alloys on the Body–Centered Cubic (bcc) Lattice

The modeling of bcc binary alloys where one species is magnetic is of interest for systems such as Fe–Al and Fe–Si [8]. Fig. 6 illustrates the two relevant stoichiometric superstructures, the FeAl structure (denoted also as B2), the $Fe_3Al$ phase (denoted also as $DO_3$), both of which can be either paramagnetic (para) of ferromagnetic (ferro). In the disordered (A2) structure, all four sublattices a, b, c, d (Fig. 6) have identical average concentrations of A and B atoms,

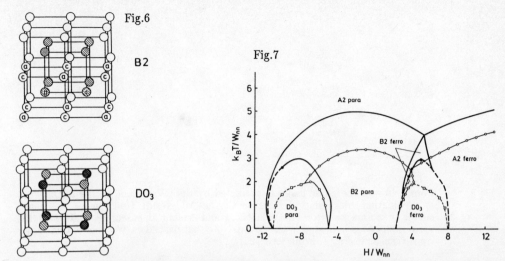

Fig.6: Bcc lattice showing the B2 structure (upper part) and DO$_3$ structure (lower part). Upper part shows assignment of four sublattices a, b, c, d. From Ref. [20].

Fig.7: Comparison of the phase diagram predicted by the Bragg–Williams mean–field approximation (MFA) (thick lines) with the phase diagram predicted by the CV method in the tetrahedron approximation (thin lines with open circles and squares), in the grand–canonical ensemble, for $W_{NNN}/W_{NN} = 0.5$, $J/W_{NN} = 0.7$. Solid dots denote tricritical points as found in the MFA; full curves denote second order transitions, broken curves first–order transitions. Nature of the various phases is indicated in the figure. From Ref. [20].

while in the B2 structure the concentrations at the b and d sublattice are the same, but differ from the concentrations at the a and c sublattices, which again are the same. In the DO$_3$ structure, the concentrations at the a and c sublattices are still the same, while the concentration at sublattice b differs from the concentration at sublattice d.

The simplest possible model treats the magnetic degree of freedom of Fe as an Ising spin $\sigma_i = \pm 1$, and hence the model treated in Ref. [20] includes contributions from nearest (NN) and next nearest (NNN) neighbors

$$\mathcal{H} = \sum_{\langle i,j \rangle_{NN}} \left\{ c_i^A c_j^A (v_{NN}^{AA} - J\sigma_i\sigma_j) + \left[ c_i^B c_j^A + c_i^A c_j^B \right] v_{NN}^{AB} + c_i^B c_j^B v_{NN}^{BB} \right\} + $$
$$+ \sum_{\langle i,j \rangle_{NNN}} \left\{ c_i^A c_j^A v_{NNN}^{AA} + \left[ c_i^B c_j^A + c_i^A c_j^B \right] v_{NNN}^{AB} + c_i^B c_j^B v_{NNN}^{BB} \right\} ; \qquad (7)$$

transforming again the $c_i^A$'s, $c_i^B$'s to pseudospins $S_i = \pm 1$ as in Eqs. (1),(2) one recognizes that three relevant couplings remain, $W_{NN} = (v_{NN}^{AA} + v_{NN}^{BB} - 2v_{NN}^{AB})/4$, $W_{NNN} = (v_{NNN}^{AA} + v_{NNN}^{BB} - 2v_{NNN}^{AB})/4$, J, and the "field" is $H = -2(v_{NN}^{AA} - v_{NN}^{BB}) - (3/2)(v_{NNN}^{AA} - v_{NNN}^{BB}) + \Delta\mu/2$. Since the A–atoms are magnetic and the B–atoms are not, there is no longer any symmetry at average concentration $\langle c \rangle = 1/2$, and due to the many phases the problem is rather complicated [20]. Again the conclusion is (Fig. 7) that different analytical approximations to the same model yield fairly different phase diagrams: in MFA there is no direct transition from A2 para to DO$_3$ para, since there is always a B2 para phase intervening, in contrast to the CV result. The former method yields reentrant phase boundaries, the latter does not, and the ordering temperatures are typically a factor 1.5 lower. Before establishing the accuracy of any of such approximations, an extension of the model for comparison with experiment [8] seems premature. In this case, the MC calculation is in fairly close agreement with the CV method, which yields a phase diagram

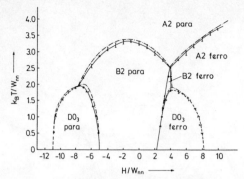

Fig.8: Comparison of the phase diagram predicted by the CV method in the tetrahedron approximation (dash–dotted lines) with the MC result (full curves or dashed curves or points or error bars), in the grand–canonical ensemble. Nature of the various phases is indicated in the figure. Note that dashed curves denote first–order transitions as in Fig. 7. From Ref. [20].

of the correct topology (Fig. 8) and overestimates the transition temperatures by a few percent only. However, more work in which the parameters of the model Eq. (7) are varied is needed before a comparison with experimental work on the Fe–Al system will become possible.

## 4. Interdiffusion in the Random Alloy (ABV) Model

We now turn to questions relating to the dynamics of ordering or unmixing in alloys. Most of the early simulations on that problem have been based on the simple but unrealistic assumption of direct interchange between A- and B–atoms on neighboring sites [35]. In reality, vacancies (V) at small concentration $c_V \ll 1$ are needed to allow interdiffusion of A and B: if an A–atom jumps (with jump rate $\Gamma_A$) to a vacant site, it leaves a vacancy on the site it left behind, to which now a neighboring B–atom may jump (with jump rate $\Gamma_B$), etc. [29,30].

Even in the simplest case where there are no interactions between particles apart from the "exclusion volume" (each lattice site can either be taken by A or by B or stay vacant, $c_i^A + c_i^B + c_i^V = 1$), there are already severe problems in understanding interdiffusion quantitatively. There has been a controversy whether there is a simple relation between the "interdiffusion constant" $D_{int}$ and the "tracer diffusion constants" $D_t^A$, $D_t^B$. Here $D_{int}$ describes how a deviation of relative concentration from equilibrium relaxes, $\delta c_k(t)/c \sim exp[-D_{int}k^2t]$, $c = <c_i^A> + <c_i^B>$, $\delta c_k(t) = \sum_i [(c_i^A - c_i^B) - (<c_i^A> - <c_i^B>)] exp[i\vec{k}\cdot\vec{r}_i]$. The tracer diffusion constants are related to the time dependence of the mean square displacement of "tagged" particles, $D_t^{A,B} = \lim_{t\to\infty} <[\vec{r}_i(t) - \vec{r}_i(0)]^2>/[2dt]$ in d dimensions, $\vec{r}_i(t)$ being the position of the tagged A–atom (or B–atom) at the lattice. According to the so–called "fast mode" theory the faster diffusing species controls interdiffusion [55]

$$D_{int} = (<c_i^B>D_t^A + <c_i^A>D_t^B)/c, \qquad (8)$$

while according to the so–called "slow mode" theory the slower diffusing species dominates [56],

$$D_{int}^{-1} = c/[<c_i^B>/D_t^A + <c_i^A>/D_t^B]. \qquad (9)$$

Both Eqs. (8), (9) are derived from the same phenomenological equations for the current densities $\vec{j}_A$, $\vec{j}_B$ of A– and B–atoms,

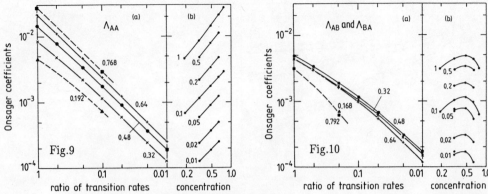

Fig.9:  Onsager coefficients $\Lambda_{AA}$ of two–dimensional alloy system plotted (a) as functions of $\Gamma_A/\Gamma_B$ with $<c_i^A>$ indicated as a parameter, and (b) as functions of $<c_i^A>$ with $\Gamma_A/\Gamma_B$ as a parameter. All curves are guides to the eye. From Ref. [41].

Fig.10:  Same as Fig. 9, but off–diagonal Onsager coefficients $\Lambda_{AB} = \Lambda_{BA}$. From Ref. [41].

$$\vec{j}_A = -(\Lambda_{AA}/k_B T)\nabla(\mu_A - \mu_V) - (\Lambda_{AB}/k_B T)\nabla(\mu_B - \mu_V) \,, \qquad (10)$$

$$\vec{j}_B = -(\Lambda_{BA}/k_B T)\nabla(\mu_A - \mu_V) - (\Lambda_{BB}/k_B T)\nabla(\mu_B - \mu_V) \,, \qquad (11)$$

where $\Lambda_{AA}$, $\Lambda_{AB}$, $\Lambda_{BA}$, $\Lambda_{BB}$ are Onsager coefficients and $\mu_A$, $\mu_B$, $\mu_V$ are chemical potentials. Both approaches neglect the nondiagonal Onsager coefficients and assume that the diagonal ones are proportional to the tracer diffusion constants, $D_t^A \sim \Lambda_{AA}$, $D_t^B \sim \Lambda_{BB}$ [55,56]. However, they differ in their assumptions about $\mu_V$: Ref. [55] assumes $\nabla\mu_V = 0$, while Ref. [56] calculates $\mu_V$ from thermodynamics (the entropy of mixing).

In view of all those assumptions, the validity of which is uncertain, a computer simulation of the random ABV model has been performed [41] where both all Onsager coefficients, all tracer diffusion coefficients and $D_{int}$ have been estimated for many choices of $<c_i^A>$, $<c_i^B>$ and the ratio $\Gamma^{-1} = \Gamma_A/\Gamma_B$. Note that Onsager coefficients can be measured by making the jump rates different in the forward and backward directions, $\Gamma_{AX} = b\Gamma_A$, $\Gamma_{A-x} = b^{-1}\Gamma_A$, which leads to a mean velocity in $+x$–direction, $v_x^{(\alpha)}$ for species of type $\alpha$, from which the Onsager coefficient follows as $\Lambda_{\alpha x}/c_\alpha = v_x^{(\alpha)}/(b - b^{-1})$ [41]. Figs. 9, 10 show that the Onsager coefficient $\Lambda_{AA}$ of the slower diffusing species for many cases is not larger than the off–diagonal ones ($\Lambda_{AB} = \Lambda_{BA}$), and hence a basic assumption of all the theories [55,56] fails! The interdiffusion constant is measured by preparing an initial state of the model where a concentration wave with wavelength $\lambda = 2\pi/k$ is present, choosing typically initial amplitudes of the waves $\delta c_A(0) = -\delta c_B(0) = 0.02$, for instance (Fig. 11). Using Eqs. (10), (11) but avoiding any further approximations one can show that in a non–interacting case [41]

$$D_{int} = \frac{\Lambda_{AA}\Lambda_{BB} - \Lambda_{AB}^2}{\Lambda_{AA} + 2\Lambda_{AB} + \Lambda_{BB}} \left[\frac{1}{<c_i^A>} + \frac{1}{<c_i^B>}\right] \,. \qquad (12)$$

Using the actually "measured" Onsager coefficients (such as shown in Fig. 9, 10) in Eq. (12) the simulation data (Fig. 11) are accounted for quantitatively. In contrast, for the case of Fig. 11, Eq. (8) would predict $D_{int} = 0.00914$, Eq. (9) $D_{int} = 0.00325$, the actual result being

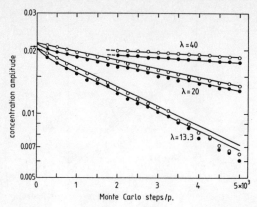

Fig.11: Amplitudes of concentration profiles as a function of time (in units of Monte Carlo steps per particle). Open circles represent A–atoms, full dots represent B–atoms, for a lattice of $80^3$ sites, $\langle c_i^A \rangle = \langle c_i^B \rangle = 0.48$, $\Gamma_A/\Gamma_B = 0.1$. Three different wavelengths $\lambda$ are shown (the arrow indicates the initial concentration amplitude for $\lambda = 40$). The curves represent the theory based on the actual Onsager coefficients. From Ref. [41]

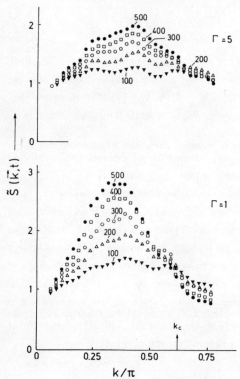

Fig.12: Structure factor $\tilde{S}(\vec{k},t)$ for $\vec{k} = (\nu,0)2\pi/L$, where $L = 80$ and $\nu$ is an integer, for a two–dimensional model quenched from infinite temperature to $k_B T/\epsilon_{AB} = 0.6$ at $\Gamma = 1$ (lower part) and $\Gamma = 5$ (upper part). The arrow shows the mean field estimate of the critical wavenumber. Various times t after the quench are shown, as indicated in the figure. Time is measured in MCS per vacancy. From Ref. [42].

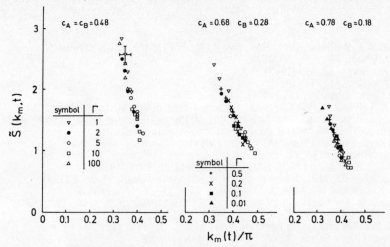

Fig.13: Plot of $S(k_m,t)$ vs. $k_m(t)$ for three different choices of concentrations as indicated. Various values of $\Gamma$ are included as shown by different symbols in the figure. From Ref. [42].

$D_{int} = 0.00565$. One finds that typically the actual $D_{int}$ is between the predictions of Eqs. (8) and (9), and there is no general rule which of these results is better. Of course, the ABV model studied there is certainly far too simplified to account for actual alloys or fluid mixtures — in the latter, hydrodynamic effects clearly will be present [57]. However, there is little reason to believe that for complicated systems such as polymer mixtures [55,56] any of the formulas Eqs. (8), (9) should be accurate.

If one introduces a repulsive energy $\epsilon_{AB}$ between A– and B–atoms into the ABV model, phase separation occurs at low enough temperatures. Quenching the system from a random initial configuration to a temperature $k_B T/\epsilon_{AB} = 0.6$ (note $k_B T_c/\epsilon_{AB} \simeq 1.07$ for $c_v = 0.04$), the homogeneous initial state is unstable and the system starts to unmix via spinodal decomposition [31–34]. This process shows up as the formation of a peak in the structure factor $S(\vec{k},t)$ (Fig. 12). Quantitatively, this behavior is the same as for the direct interchange model [35]. The linearized Cahn [58] theory predicts $k_c$ {where initially $S(\vec{k},t)$ is independent of time} roughly correctly $\{k_c = arccos[k_B T/\epsilon_{AB}(1 - \Phi_v) - 1]\}$ [42], although the growth is much slower than exponential. It is also interesting that on a plot of the structure factor $S(k_m,t)$ versus its peak positions $k_m(t)$ different choices of $\Gamma$ superimpose on a master curve (Fig. 13). This implies that the $\Gamma$–dependence can be absorbed in a redefinition of the time scale, and explains why different systems behave qualitatively in a similar manner.

5. Discussion

This contribution has described some applications of the Monte Carlo method to clarify cooperative phenomena in metallic alloys. It has been shown that approximate theories for both static and dynamic properties can be stringently tested, and valuable hints for the improvement of such analytical theories are gained. At the same time, one can test to what extent simple models actually represent physical properties of real materials faithfully. Since so far mostly extremely simplified cases have been studied, and much more work on more realistic models seem feasible, we expect that this approach will be become a valuable tool for investigating problems in materials science in the near future.

Acknowledgments: Part of the results described here have been obtained jointly with B. Dünweg [20], K. Kehr and S.M. Reulein [41] and K. Yaldram [42]. It is a great pleasure to thank them for a very fruitful collaboration.

# References

1. W.L. Bragg and E.J. Williams, Proc. Roy. Soc. A145, 699 (1934); A152, 231 (1935)
2. W. Shockley, J. Chem. Phys. 6, 130 (1938)
3. Y.Y. Li, J. Chem. Phys. 17, 449 (1949)
4. R. Kikuchi, Phys. Rev. 81, 998 (1951)
5. R. Kikuchi, J. Chem. Phys. 60, 107 (1974)
6. N.S. Golosov, L.E. Popov and L.Y. Pudan, J. Phys. Chem. Solids 34, 1149, 1159 (1973)
7. D. De Fontaine and R. Kikuchi, in Applications of Phase Diagrams in Metallurgy and Ceramics, Vol.2, p.976 (NBS Special Publication 496, Washington 1978)
8. S.V. Semenovskaya, Phys. stat. sol. (b) 64, 291 (1974)
9. K. Binder, Phys. Rev. Lett. 45, 811 (1980)
10. M.K. Phani, J.L. Lebowitz and M.H. Kalos, Phys. Rev. B21, 4027 (1980)
11. K. Binder, J.L. Lebowitz, M.K. Phani and M.H. Kalos, Acta Metall. 29, 1655 (1981)
12. K. Binder, W. Kinzel and W. Selke, J. Magn. Mag. Mat. 31–34, 1145 (1983)
13. J.M Sanchez and D. De Fontaine, Phys. Rev. B21, 216 (1980); B25, 1759 (1982); J.M. Sanchez, D. De Fontaine and W. Teitler, Phys. Rev. B26, 1465 (1982)
14. T. Mohri, J.M. Sanchez and D. De Fontaine, Acta Metall. 33, 1171 (1985)
15. U. Gahn, J. Phys. Chem. Solids 43, 977 (1982); 47, 1153 (1986)
16. A. Finel and F. Ducastelle, Europhys. Lett. 1, 135 (1986)
17. H.T. Diep, A. Ghazali, B. Berge and P. Lallemand, Europhys. Lett. 2, 603 (1986)
18. G. Inden, Acta Metall. 22, 945 (1974); Z. Metallkde. 66, 577, 648 (1975)
19. H. Ackermann, S. Crusius and G. Inden, Acta Metall. 34, 2311 (1986)
20. B. Dünweg and K. Binder, Phys. Rev. B36, 6935 (1987)
21. For reviews, see D. De Fontaine, in Solid State Physics (ed. by H. Ehrenreich, F. Seitz and D. Turnbull) Vol. 34, p.73 (Academic Press, New York 1979), and Refs. 22–24
22. A.G. Khachaturyan, phys. stat. sol. (b) 60, 9 (1973); Theory of Structural Transformations in Solids (J. Wiley & Sons, New York 1983)
23. K. Binder, in Festkörperprobleme (Advances in Solid State Physics), Vol.26, p.133, P. Grosse (ed.) (Vieweg, Braunschweig 1986)
24. A. Gonis and L.M. Stocks (eds.) Alloy Phase Stability (Kluwer Acad. Publishers, Dordrecht 1989)
25. J. Als–Nielsen, in Phase Transitions and Critical Phenomena, Vol.5a (C. Domb and M.S. Green, eds.) p.88 (Academic Press, New York 1976)
26. K. Binder (ed.) Monte Carlo Methods in Statistical Physics (Springer, Berlin 1979)
27. K. Binder (ed.) Applications of the Monte Carlo Method in Statistical Physics (Springer, Berlin 1984)
28. K. Binder and D.W. Heermann, Monte Carlo Simulation in Statistical Physics: An Introduction (Springer, Berlin 1988)
29. J.R. Manning, Diffusion Kinetics for Atoms in Crystals (Van Nostrand, Princeton 1968)
30. C.P. Flynn, Point Defects and Diffusion (Clarendon Press, Oxford 1972)
31. For reviews, see J.D. Gunton, M. San Miguel and P.S. Sahni, in Phase Transitions and Critical Phenomena, Vol.8 (C. Domb and J.L. Lebowitz, eds.) p.267 (Academic Press, New York 1987)
32. K. Binder, Repts. Progr. Phys. 50, 783 (1987)
33. K. Binder, in Ref. 24, p.233
34. K. Binder, in Materials Science and Technology, Vol.5: Phase Transformations in Materials (P. Haasen, ed.9 Chapter 7 (VCH Verlagsgesellschaft, Weinheim, in press)
35. For reviews, see Refs. 31–34 and K. Binder and M.H. Kalos, in Ref. 26, p.225
36. For reviews, see K.W. Kehr and K. Binder, in Ref. 27, p.181, and Ref. 37
37. G.E. Murch, in Diffusion in Solids II (G.E. Murch, N.S. Nowick, eds.) (Academic Press, New York 1984)
38. K.W. Kehr, K. Kutner and K. Binder, Phys. Rev. B23, 4931 (1981)
39. R. Kutner, K. Binder and K.W. Kehr, Phys. Rev. B26. 2967 (1982)
40. R. Kutner, K. Binder and K.W. Kehr, Phys. Rev. B28, 1846 (1983)
41. K.W. Kehr, K. Binder and S.M. Reulein, Phys. Rev. B39, 4891 (1989)
42. K. Yaldram and K. Binder, preprints
43. M. Hansen, Constitution of Binary Alloys (McGraw-Hill, New York 1958)
44. J. Hafner, From Hamiltonians to Phase Diagrams: The Electronic and Statistical Mechanical Theory of sp–Bonded Metals and Alloys (Springer, Berlin 1985)
45. R.C. Kittler and L.M. Falicov, Phys. Rev. B18, 2506 (1978); B19, 291 (1979)

46. W. Schweika and J.G. Haubold, in <u>Atomic Transport and Defects in Metals by Neutron Scattering</u> (C. Janot, T. Petry, D. Richter and T. Springer, eds.) p.22 (Springer, Berlin 1986); Phys. Rev. <u>B37</u>, 9240 (1988)
47. J.M. Cowley, Phys. Rev. <u>77</u>, 669 (1950)
48. C. Sykes and F.W. Jones, Proc. Roy. Soc. London, Sec. <u>A157</u>, 213 (1938); R.L. Orr, Acta Metall. <u>8</u>, 489 (1960); R.L. Orr. K. Luciat–Labry and R. Hultgren. Acta Metall. <u>8</u>, 431 (1960)
49. D.T. Keating and B.E. Warren, J. Appl. Phys. <u>22</u>, 286 (1951)
50. S.C. Moss, J. Appl. Phys. <u>35</u>, 3547 (1964)
51. J.M. Cowley, J. Appl. Phys. <u>21</u>, 24 (1950)
52. L.H. Schwartz and J.B. Cohen, J. Appl. Phys. <u>36</u>, 598 (1965)
53. P. Bardham and J.B. Cohen, Acta Cryst. <u>A32</u>, 597 (1976)
54. J.L. Lebowitz, M.K. Phani and D.F. Styer, J. Stat. Phys. <u>38</u>, 413 (1985)
55. E.J. Kramer, P. Green and C.J. Palmstrom, Polymer <u>25</u>, 473 (1984)
56. K. Binder, J. Chem. Phys. <u>79</u>, 6387 (1983); F. Brochard, J. Jouffray and P. Levinson, Macromolecules <u>16</u>, 1638 (1983); K. Binder, Colloid Polymer Sci. <u>265</u>, 273 (1987)
57. W. Hess and H.L. Frisch, Europhys. Lett. <u>5</u>, 391 (1988); W. Hess and A.Z. Akcasu, J. Phys. (Paris) <u>49</u>, 1261 (1988)
58. J.W. Cahn, Acta Metall. <u>9</u>, 795 (1961); <u>14</u>, 1685 (1966); Trans. Metall. Soc. AIME <u>242</u>, 166 (1968)

# Monte Carlo Analysis of the Ising Model and CAM

*N. Ito and M. Suzuki*

Department of Physics, Faculty of Science, University of Tokyo,
Hongo, Bunkyo-ku, Tokyo 113, Japan

**Abstract** A cluster–effective–field approximation, namely a modified Bethe approximation, is proposed which is canonical in the CAM sense. This is easier to treat theoretically and to calculate explicitly than the previous cluster–Bethe approximation. The Weiss and modified Bethe approximations for the Ising ferromagnets on the square and simple cubic lattices are solved by the Monte Carlo method and the CAM coefficients of susceptibility and magnetization are obtained. We have estimated the critical point and critical exponent of susceptibility with good precision. It is observed that the cluster–Weiss approximation does not show good CAM scaling. It produces misleading values of exponents from the CAM coefficients of the clusters we have calculated. Our modified Bethe approximation can be easily extended to the super–effective–field theory and will be useful for studying exotic phase transitions.

## 1  Introduction

The coherent–anomaly method (CAM)[1]–[13] has cast light on the classical approximations of critical phenomena. If some series of such approximations satisfies the degree–of–approximation scaling (CAM scaling), it is called *canonical*. Thus it becomes important to construct a good canonical approximation for the study of critical phenomena. Many canonical approximations have been proposed, for example, the cluster–effective–field approximations[2]–[6], [14], perturbation–series[15]–[16] and cluster–variation method[17]. With these approximations and the CAM, many systems have been studied, that is, the Ising models[2]–[6], [14], [17], [18], [19], quantum Heisenberg model[20], [21], Potts model[22], spin–glass[23], self–avoiding walk[24], percolation models[25]–[26] and contact processes[27]. The mechanism of the CAM has been studied and justified in a phenomenological sense and partially in a mathematical sense[2].

The calculations of cluster–effective–field approximations are not easy when the clusters become fairly large. Analytic calculations are possible only for small clusters of simple models, although one can treat analytically rather large lattices for two–dimensional non–random Ising models using the transfer–matrix technique[3], [14]. The Monte Carlo simulation will be powerful when we try to calculate larger lattices and to obtain better estimations. In the present paper, the cluster–effective–field approximations of the two– and three–dimensional Ising ferromagnets are studied using the Monte Carlo method. The purpose of the present study is to test the CAM scaling for the cluster effective field approximations and get quantitative estimations for the three–dimensional ferromagnetic Ising model. We have tried two approximations, namely the Weiss–like and Bethe–like ones. The effective field which represents the cooperative effect of spins which are in the outside of the cluster is applied on the boundary spins of the cluster with weights equal to the numbers of free bonds at each spin. In the ordinary Bethe–like case, the self–consistency condition is that the average of the center spin is equal to that of each boundary spin induced by effective fields. In the present Bethe–like case, the average of the center spin is assumed to be equal to the average of the boundary magnetization per boundary spin. We call this the *modified-Bethe approximation*. In the previously proposed cluster–Bethe approximation[2], effective fields may be different at each site if the boundary sites are not equivalent under the action of the symmetry group of the relevant cluster. This approximation is better for small clusters, which allow an analytic treatment. When the cluster becomes larger, however, the number of different fields becomes enormous and consequently the calculations may become very complicated. Therefore our modified Bethe approximation will be practical. As is shown in the following sections, it has a simple mathematical structure and its degree–of–approximation is better than the Weiss approximation. It is also observed that the critical point of this approximation gives very good upper–bounds of the true critical point.

Monte Carlo estimations have become much more precise than before, because the simulation speed has grown exponentially in time[28] and will keep this growth rate for the time being[28]. Ito and Kanada have proposed effective algorithms for the Ising Monte Carlo on the vector processor[28], [29]. Their source codes can update $0.93 \times 10^9$ spins a second on the vector processor HITAC S820/80, which is a single processor machine whose peak performance is 3G FLOPS. In this sense the Monte Carlo analysis of cluster effective–field approximations will become more and more important. But the Monte Carlo estimations are not so precise as analytic estimations. Thus we have to be careful not to spoil final results by large errors and it is crucial how the Monte Carlo data are analyzed.

For ferro– or antiferro–magnetic phase transitions, the effective–field theory is easy to construct. It is not straightforward, however, for *exotic* phase transitions such as the spin–glass and chiral–order transitions. The super–effective–field theory (SEFT) is a promising idea for these problems[30]–[35]; it has been applied successfully to the chiral–order transition[36]. The SEFT makes use of the Bethe–like self–consistency condition and our modified–Bethe approximation can be applied to the SEFT.

The cluster Weiss and modified–Bethe approximations are described in the next section. The expressions for CAM coefficients are also given. They are obtained using the Monte Carlo methods in the third section. The results are analyzed and the true critical point and exponents are estimated in the fourth section. The final section contains the summary and discussion.

## 2 Cluster Approximations

The Hamiltonian of the Ising model which includes the thermal factor $(-\beta)$ is given by

$$H = K \sum_{\substack{|i-j|=1 \\ i,j \in \Lambda}} \sigma_i \sigma_j + L \sum_{i \in \Lambda} \sigma_i \quad (\sigma_i = \pm 1, \quad K = \beta J \quad \text{and} \quad L = \beta h), \tag{2.1}$$

where $\Lambda$ denotes the vertex set of the relevant lattice and $J$ and $h$ denote the coupling constant and the uniform external field, respectively. What we want to know here is the behavior of this model in the thermodynamic limit. It is impossible in most cases to calculate the free energy in the thermodynamic limit, that is,

$$f = \lim_{\Lambda \to \infty} \frac{1}{|\Lambda|} \log \mathrm{Tr}_\Lambda e^H, \qquad (2.2)$$

where $|\Lambda|$ denotes the number of elements in the set $\Lambda$. A finite cluster in $\Lambda$ is denoted by $\Omega$. Instead of the full trace in (2.2), the effective density operator and the effective Hamiltonian of $\Omega$ defined by

$$\rho(\{\sigma\}_\Omega) = \mathrm{Tr}_{\Lambda-\Omega} e^H = e^{H_{\mathrm{eff}} + f \cdot |\Lambda - \Omega|} \qquad (2.3)$$

are considered in order to study the thermodynamic properties of a finite cluster $\Omega$ in an infinite lattice. In (2.3), $\mathrm{Tr}_{\Lambda-\Omega}$ denotes the trace operation for the spins in $\Lambda - \Omega$. The Hamiltonian of the Ising model is a local and classical quantity. Therefore the effective Hamiltonian of $\Omega$ can be expressed as

$$H_{\mathrm{eff}} = H_\Omega + H_{\mathrm{eff}}(\partial \Omega). \qquad (2.4)$$

where $H_\Omega$ is the original Hamiltonian of the cluster $\Omega$ and $H_{\mathrm{eff}}(\partial\Omega)$ is some function of the boundary spins of the cluster $\Omega$, which is defined by (2.3). The $H_{\mathrm{eff}}(\partial\Omega)$ is invariant under the action of the symmetry group of the cluster $\Omega$, because the phase of ferromagnets is spatially uniform (we do not consider here the roughening phase). The effective fields on the boundary are introduced in this manner.

The coupling constants in the $H_{\mathrm{eff}}(\partial\Omega)$ are complicated functions of temperature and external magnetic field and it is difficult to obtain the exact form of these functions for a finite cluster. Thus we consider only one–point effective fields and neglect all other effective interactions. Then the value of effective fields can be determined from the self–consistency conditions. This is the essential point of the cluster effective–field approximations.

The physical quantities show classical singularities in the effective field approximations. For example, the spontaneous magnetization $m_s$ and susceptibility $\chi$ behave near the critical point of the approximation like

$$\lim_{K \to +0} \frac{m_s}{\bar{m}\sqrt{(K - K_c)/K_c}} = 1 \qquad (2.5)$$

and

$$\lim_{K \to -0} \frac{\chi}{\bar{\chi} K_c/(K_c - K)} = 1. \qquad (2.6)$$

Therefore the $m_s$ and $\chi$ of the effective field approximation have the classical exponents, 1/2 and 1, respectively.

The amplitude of the most singular term is called the CAM coefficient, which contains enough information to estimate the true singularity, not the classical value. In the above notation, $\bar{m}$ and $\bar{\chi}$ are the CAM coefficients of the spontaneous magnetization and susceptibility.

Two kinds of approximations are formulated in the following. One is the Weiss–type cluster approximation and the other is the modified Bethe–type cluster approximation.

## 2.1 Weiss Approximation

The cluster Weiss approximation[3] is reviewed before we describe our modified Bethe approximation. This is not an effective–field theory in the above sense, because the self–consistency does not directly claim the translational invariance of the one–point function. This approximation has been studied for smaller clusters and its nature has been studied better than that of the Bethe approximation. Thus we can compare here our Monte Carlo results with the previous results and with our modified Bethe approximations. For these reasons, the Weiss approximation is treated here. The following expressions are essentially the same as in [3], though we will introduce our simplified notations.

This approximation treats the following Hamiltonian:

$$H_{\text{Weiss}} = KA + \frac{K}{J}hB + Kh_{\text{eff}}C, \tag{2.7}$$

where

$$A = \sum_{\substack{|i-j|=1 \\ i,j \in \Omega}} \sigma_i \sigma_j, \quad B = \sum_{i \in \Omega} \sigma_i \quad \text{and} \quad C = \sum_{i \in \partial\Omega} w_i \sigma_i. \tag{2.8}$$

In the last definition, $w_i$ denotes the number of free bonds at the $i$-site, that is, the number of the sites which are the nearest neighbor of $i$ and do not belong to the cluster $\Omega$. The parameter $h$ is a uniform magnetic field and $h_{\text{eff}}$ is the mean–field at the boundary.

The self–consistency condition in the Weiss approximation is

$$<\sigma_0> = h_{\text{eff}}, \tag{2.9}$$

where this $<\cdot>$ denotes the thermal average with the Hamiltonian (2.7). This $<\sigma_0>$ is expanded in $h$ and $h_{\text{eff}}$ to discuss the behavior near the critical point:

$$<\sigma_0> = \sum_{i,j=0}^{\infty} a_{ij} h^i h_{\text{eff}}^j, \tag{2.10}$$

where

$$a_{ij} = \frac{1}{i!j!}\left[\frac{\partial^{i+j} <\sigma_0>}{\partial h^i \partial h_{\text{eff}}^j}\right]_{h=h_{\text{eff}}=0}, \tag{2.11}$$

which is zero if $(i+j)$ is even.

The critical point, $K_c$, is determined by the equation

$$a_{01} - 1 = 0. \tag{2.12}$$

It is known that the critical temperature of this approximation is always higher than the true critical temperature[2].

The skeletonized spontaneous magnetization $m_s$ in the ordered phase $(K > K_c)$ is

$$m_s = <\sigma_0> = h_{\text{eff}} = \sqrt{\frac{1-a_{01}}{a_{03}}}, \tag{2.13}$$

which is obtained from the equation

$$a_{01} h_{\text{eff}} + a_{03} h_{\text{eff}}^3 = 0. \tag{2.14}$$

Therefore the CAM coefficient of magnetization $\bar{m}_{\text{Weiss}}$ is given by

$$\bar{m}_{\text{Weiss}} = \sqrt{\frac{bK_c}{a_{03}}}, \tag{2.15}$$

where $K - K_c$ is used as the reduced temperature and

$$b = \left[\frac{\partial(1 - a_{01})}{\partial K}\right]_{K=K_c, h=h_{\text{eff}}=0}.\tag{2.16}$$

The skeletonized susceptibility in the disordered phase ($K < K_c$) is obtained from the equation

$$\chi = \left[\frac{d < \sigma_0 >}{dh}\right]_{h=0} = \frac{a_{10}}{1 - a_{01}},\tag{2.17}$$

which is obtained from the equation

$$a_{01}h + a_{10}h_{\text{eff}} = h_{\text{eff}}.\tag{2.18}$$

Thus the CAM coefficient $\bar{\chi}_{\text{Weiss}}$ is given by

$$\bar{\chi}_{\text{Weiss}} = -\frac{a_{10}}{bK_c}.\tag{2.19}$$

The expressions for $a_{ij}$ and $b$ are obtained from (2.11) and (2.16) using the following well-known formula:

$$\frac{\partial < Q >}{\partial x} = < Q\frac{\partial H}{\partial x} > - < Q >< \frac{\partial H}{\partial x} >,\tag{2.20}$$

where $H$, $x$ and $Q$ are the Hamiltonian of this ensemble, a parameter in the Hamiltonian and some operator which does not depend on $x$, respectively. If the Hamiltonian does not contains $x^2$ or higher terms, this formula can be applied repeatedly. The parameters of our Hamiltonian (2.7) are $h$, $h_{\text{eff}}$ and $K$ and

$$\frac{\partial H}{\partial h} = \frac{K}{J}B, \quad \frac{\partial H}{\partial h_{\text{eff}}} = KC \quad \text{and} \quad \frac{\partial H}{\partial K} = A + \frac{h}{J}B + h_{\text{eff}}C.\tag{2.21}$$

With the help of the above formulas (2.21), the expressions for $a_{01}$, $a_{03}$, $a_{10}$ and $b$ are obtained after some simple calculations. They are given by

$$a_{01} = K < \sigma_0 C >_0,\tag{2.22}$$

$$a_{03} = K^3\{< \sigma_0 C^3 >_0 - 3 < \sigma_0 C >_0 < C^2 >_0\},\tag{2.23}$$

$$a_{10} = \beta < \sigma_0 B >_0 = \frac{K}{J} < \sigma_0 B >\tag{2.24}$$

and

$$b = - < \sigma_0 C >_0 - K\{< \sigma_0 CA >_0 - < \sigma_0 C >_0 < A >_0\},\tag{2.25}$$

where $< \cdot >_0$ denotes the thermal average without magnetic field and mean field at the critical point determined from (2.12), that is, $h = h_{\text{eff}} = 0$ and $K = K_c$. Thus the necessary thermal expectation values are

$$< \sigma_0 C >_0, \quad < \sigma_0 C^3 >_0, \quad < C^2 >_0, \quad < A >_0, \quad < \sigma_0 CA >_0 \quad \text{and} \quad < \sigma_0 B >_0.\tag{2.26}$$

They are all expressed as some kinds of summations of correlations in a finite cluster with free boundary conditions.

## 2.2 Modified–Bethe Approximation

Our cluster modified–Bethe approximation treats the same form of Hamiltonian as that of the Weiss approximation, namely

$$H_{\text{m-Bethe}} = KA + \frac{K}{J}hB + Kh_{\text{eff}}C, \tag{2.27}$$

where $A$, $B$ and $C$ are defined in (2.8). The self–consistency condition here is, however, different from the previous one. It requires that the expectation value of the central spin is equal to the average of the expectation values of the boundary spins, namely,

$$<\sigma_0> = \frac{1}{|\partial\Omega|}<D>, \tag{2.28}$$

where $D$ is defined by

$$D = \sum_{i\in\partial\Omega}\sigma_i, \tag{2.29}$$

which is the boundary magnetization. The self–consistency condition claims that the one–point function at the center is the same as those at the boundary on average. Both sides of the equation (2.28) are expanded in $h$ and $h_{\text{eff}}$. The LHS is the same as (2.10). The RHS is

$$\frac{1}{|\partial\Omega|}<D> = \sum_{i,j=0}^{\infty} b_{ij} h^i h^j_{\text{eff}}, \tag{2.30}$$

where

$$b_{ij} = \frac{1}{i!j!}\frac{1}{|\partial\Omega|}\left[\frac{\partial^{i+j}<D>}{\partial h^i \partial h^j_{\text{eff}}}\right]_{h=h_{\text{eff}}=0}, \tag{2.31}$$

which is zero if $(i+j)$ is even.

Performing almost the same procedures as in the Weiss approximation, we derive the expressions of our approximation as follows. The critical point is determined by the equation

$$a_{01} - b_{01} = 0. \tag{2.32}$$

The CAM coefficients are

$$\bar{m}_{\text{m-Bethe}} = a_{01}\sqrt{-\frac{wK_c}{a_{03} - b_{03}}} \tag{2.33}$$

and

$$\bar{\chi}_{\text{m-Bethe}} = -\frac{a_{01}b_{10} - a_{10}b_{01}}{wK_c}, \tag{2.34}$$

where $w$ is

$$w = \left[\frac{\partial(a_{01} - b_{01})}{\partial K}\right]_{K=K_c, h=h_{\text{eff}}=0}. \tag{2.35}$$

The expressions for $a_{ij}$ have the same forms as those of the Weiss approximation, though the critical temperature at which these expressions should be evaluated is different and is determined by (2.32). The coefficients $\{b_{ij}\}$ are obtained by replacing the quantity $\sigma_0$ by the variable $D/|\partial\Omega|$. Therefore we have

$$b_{01} = K<DC>_0 / |\partial\Omega|, \tag{2.36}$$

$$b_{03} = K^3\{<DC^3>_0 - 3<DC>_0<C^2>_0\}/|\partial\Omega|, \tag{2.37}$$

and
$$b_{10} = \beta <DB>_0 / |\partial\Omega| \qquad (2.38)$$

$$w = \quad K\{<\sigma_0 CA>_0 - <\sigma_0 C>_0 <A>_0\} + <\sigma_0 C>_0$$
$$-(K\{<DCA>_0 - <DC>_0 <A>_0\} + <DC>_0)/|\partial\Omega|. \qquad (2.39)$$

Thus our necessary thermal expectation values are

$$<DC>_0, \quad <DC^3>_0, \quad <DCA>_0 \quad \text{and} \quad <DB>_0 \qquad (2.40)$$

together with (2.26).

## 3 Monte Carlo Estimation of the CAM coefficients

The critical points and the CAM coefficients of the two- and three-dimensional Ising models are calculated with Monte Carlo simulations. Circle or sphere shapes are used which are defined by the set of lattice points

$$\Omega_R = \{x \in \Lambda : |x| \leq r(R)\}, \quad r(R) = \frac{R-1}{2} + 0.01, \qquad (3.1)$$

where $R$ is an odd integer, which we call the diameter of the cluster $\Omega_R$, and the lattice constant is assumed to be unity.

For each cluster and the corresponding approximation, the critical points have to be estimated and then the CAM coefficients are estimated at these temperatures. The LHS of (2.12) and (2.32), $\bar{\chi}$ and $\bar{m}$ are calculated at several temperatures with small Monte Carlo steps and the rough estimations of the critical points are obtained. Then they are calculated at several temperatures (about 16 points) around the critical point with large Monte Carlo steps. The critical point and the CAM coefficients are determined by least-squares fitting of linear functions for the results as functions of the temperature. Fortunately, the critical points do not become much worse because they include only the two-point functions and the behavior of the LHS functions of (2.12) and (2.32) becomes steeper as the cluster becomes larger.

The heat bath algorithm is used for the Monte Carlo simulation. The same two-sublattice flip method, 32-multispin coding technique and the vectorized random-number generation are used as in Refs. [28], [29]. The logical operations for spin-flip are, however, different from those in Ref. [28], [29], because the heat bath algorithm is used here and the cluster has free boundary conditions while the previous logical operations were for the Metropolis one and the boundary was periodic.

The obtained results of the critical points and the CAM coefficients are listed in Tables 1–4. The results for square lattice are shown in Table 1 for the Weiss approximation and Table 2 for the modified Bethe approximation. For cubic lattices, Table 3 shows the Weiss approximation and Table 4 shows the modified Bethe one. The Monte Carlo data used to get the results in Tables 1–4 are shown in Tables 5 and 6. The data from a single Monte Carlo run are divided into several pieces (five to ten) and some steps omitted from the edges of pieces to avoid correlations. The sums of correlations are averaged in each piece and the errors are estimated from the deviation in each piece. These steps are sufficiently large compared with the previously known

Table 1: The results of two–dimensional Weiss approximations

| $R$ | $1/K_c$ | $\bar{m}$ | $\bar{\chi}$ |
|---|---|---|---|
| 3 | 3.415732 | 0.754422 | 0.321861 |
| 5 | 3.0931 ± 0.0014 | 0.8090 ± 0.0032 | 0.40409 ± 0.00033 |
| 11 | 2.66305 ± 0.00062 | 0.9444 ± 0.0056 | 0.6627 ± 0.0012 |
| 21 | 2.49624 ± 0.00023 | 1.0544 ± 0.0063 | 0.9413 ± 0.0016 |
| 31 | 2.43268 ± 0.00017 | 1.1180 ± 0.0092 | 1.1670 ± 0.0023 |
| 41 | 2.39844 ± 0.00019 | 1.176 ± 0.012 | 1.3578 ± 0.0042 |
| 51 | 2.37663 ± 0.00017 | 1.202 ± 0.022 | 1.5305 ± 0.0072 |
| 61 | 2.36165 ± 0.00017 | 1.253 ± 0.026 | 1.6957 ± 0.0080 |
| 81 | 2.34166 ± 0.00020 | 1.302 ± 0.065 | 1.997 ± 0.019 |
| 101 | 2.32939 ± 0.00023 | 1.254 ± 0.061 | 2.250 ± 0.027 |

Table 2: The results of two–dimensional modified Bethe approximations

| $R$ | $1/K_c$ | $\bar{m}$ | $\bar{\chi}$ |
|---|---|---|---|
| 3 | 2.88542 | 0.961233 | 0.5000 |
| 5 | 2.6989 ± 0.0017 | 1.0886 ± 0.0033 | 0.65001 ± 0.00038 |
| 11 | 2.46258 ± 0.00089 | 1.456 ± 0.010 | 1.1531 ± 0.0025 |
| 21 | 2.38235 ± 0.00020 | 1.7601 ± 0.0058 | 1.7138 ± 0.0026 |
| 31 | 2.35158 ± 0.00014 | 1.9624 ± 0.0095 | 2.1692 ± 0.0039 |
| 41 | 2.33486 ± 0.00012 | 2.111 ± 0.010 | 2.5689 ± 0.0052 |
| 51 | 2.32405 ± 0.00013 | 2.268 ± 0.014 | 2.9193 ± 0.0063 |
| 61 | 2.316914 ± 0.000085 | 2.347 ± 0.022 | 3.2348 ± 0.0091 |
| 81 | 2.30714 ± 0.00019 | 2.564 ± 0.040 | 3.837 ± 0.024 |
| 101 | 2.30101 ± 0.00016 | 2.769 ± 0.045 | 4.369 ± 0.027 |

Table 3: The results of three–dimensional Weiss approximations

| $R$ | $1/K_c$ | $\bar{m}$ | $\bar{\chi}$ |
|---|---|---|---|
| 3 | 5.44678 | 0.7381737 | 0.193942 |
| 5 | 5.07427 ± 0.00069 | 0.7923 ± 0.018 | 0.23055 ± 0.00013 |
| 11 | 4.72118 ± 0.00073 | 0.904 ± 0.016 | 0.30240 ± 0.00091 |
| 21 | 4.60268 ± 0.00052 | 0.877 ± 0.075 | 0.3713 ± 0.0029 |
| 25 | 4.58407 ± 0.00053 | | 0.3895 ± 0.0036 |
| 31 | 4.56588 ± 0.00039 | | 0.4220 ± 0.0054 |
| 41 | 4.54841 ± 0.00037 | | 0.4461 ± 0.0069 |

relaxation times[37]–[39]. Therefore the error estimations are expected to have worked correctly. The errors are those of $1\sigma$. The results of $\bar{m}$ have large errors associated with them, especially in the three–dimensional case. We cannot obtain some of them and their places are left blank. The $K_c$ and $\bar{\chi}$ of the Weiss approximation for smaller clusters are calculated analytically in [3] and those results and our present estimations are consistent with each other for such smaller clusters. The results of $K_c$ and $\bar{\chi}$ of the Weiss and our modified Bethe approximations are shown in Figs. 1 and 2, respectively.

Table 4: The results of three-dimensional Bethe approximations

| $R$ | $1/Kc$ | $\bar{m}$ | $\bar{\chi}$ |
|---|---|---|---|
| 3 | 4.9326070 | 0.9203280 | 0.25 |
| 5 | 4.7473 ± 0.0017 | 0.9380 ± 0.0063 | 0.29720 ± 0.00039 |
| 11 | 4.59989 ± 0.00075 | 1.090 ± 0.017 | 0.3891 ± 0.0011 |
| 21 | 4.55178 ± 0.00043 | 1.207 ± 0.092 | 0.4798 ± 0.0026 |
| 25 | 4.54343 ± 0.00047 | | 0.5003 ± 0.0040 |
| 31 | 4.53598 ± 0.00041 | | 0.5383 ± 0.0054 |
| 41 | 4.52941 ± 0.00041 | | 0.589 ± 0.012 |

Table 5: The Monte Carlo steps for the two-dimensional lattices shown in Tables 1 and 2. The results of $R = 3$ are analytic.

| $R$ | Total Monte Carlo steps |
|---|---|
| 3 | Analytic calculation |
| 5 | $1.1 \times 10^7$ |
| 11 | $1.1 \times 10^7$ |
| 21 | $1.0 \times 10^8$ |
| 31 | $1.0 \times 10^8$ |
| 41 | $1.0 \times 10^8$ |
| 51 | $8.1 \times 10^7$ |
| 61 | $1.0 \times 10^8$ |
| 81 | $1.0 \times 10^8$ |
| 101 | $1.0 \times 10^8$ |

Table 6: The Monte Carlo steps for the three-dimensional lattices shown in Tables 3 and 4. The results of $R = 3$ are analytic.

| $R$ | Total Monte Carlo steps |
|---|---|
| 3 | Analytic calculation |
| 5 | $1.0 \times 10^8$ |
| 11 | $1.0 \times 10^8$ |
| 21 | $1.0 \times 10^8$ |
| 25 | $1.0 \times 10^8$ |
| 31 | $1.0 \times 10^8$ |
| 41 | $5.0 \times 10^7$ |

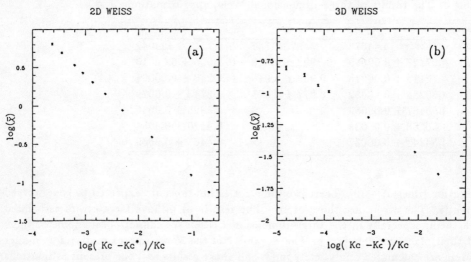

Figure 1: The results of $\bar{\chi}$ of the cluster-Weiss approximations are plotted. The two-dimensional results are shown in (a) and the three-dimensional ones are in (b).

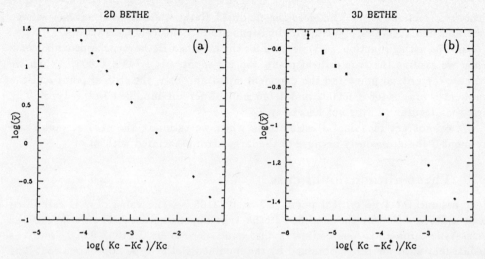

Figure 2: The results of $\bar{\chi}$ of the modified–Bethe approximations are shown: (a) is the two–dimensional and (b) is the three–dimensional case.

## 4 CAM Analysis

The results in the previous sections are analyzed by using the CAM scaling[1] and we try to estimate the true critical points and exponents. In the scaling region, the CAM coefficients behave as

$$\bar{\chi} \sim (K_c^* - K_c)^{-\psi}, \quad \gamma = 1 + \psi \tag{4.1}$$

and

$$\bar{m} \sim (K_c - K_c^*)^{\phi}, \quad \beta = \frac{1}{2} - \phi, \tag{4.2}$$

where $K_c^*$, $\gamma$ and $\beta$ denote the true critical point, the exponents of susceptibility and magnetization, respectively.

In two dimensions, these values are known, that is, $K_c^* = 0.440687\cdots$, $\gamma = 7/4$ and $\beta = 1/8$[40]. In three dimensions, they have been estimated by many authors in several methods. The obtained critical points are consistent with $K_c^* = 0.22165$[41]–[44]. There is slight inconsistency in the exponents. The value of $\gamma$ is about 1.24–1.25[45], [46]. The value of $\beta$ is estimated to be about 0.313–0.325[43], [47], [48]. We have assumed that the corrections to the scaling (4.1) and (4.2) are

$$f = \delta K_c^\lambda (a_0 + a_1 \delta K_c) \quad \delta K_c = K_c^* - K_c. \tag{4.3}$$

The errors of critical temperatures are ignored in the following fittings, because they are small compared with those of CAM coefficients.

### 4.1 Two–dimensional case

It is easily observed that the Weiss–type approximation does not show good CAM scaling behavior. When we use the true critical point, $K_c^* = 0.440687\cdots$, the values of $\gamma$ and $\beta$ are estimated to be 1.66 and 0.32 for the Weiss–type approximations and 1.73 and 0.14 for the modified Bethe–type approximation from the fitting of the function without correction term. Therefore the modified Bethe–type approximations will be preferable and the Weiss–type approximations produce misleading estimates.

Now the fitting function, (4.3), is tried for the modified Bethe–type approximations. When we assume the true critical point, the exponent $\gamma$ is $1.745 \pm 0.003$. When we estimate the critical point and the exponent simultaneously, the critical point and the exponent $\gamma$ are $0.4408 \pm 0.0002$ and $1.745 \pm 0.007$ for the modified Bethe–type. The correction term in (4.3) is not necessary.

We cannot get meaningful estimations when we estimate the critical point and exponent $\beta$ simultaneously because of the large error associated with $\bar{m}$.

## 4.2 Three–dimensional case

If we assume the true critical point, $K_c^* = 0.22165\cdots$, the value of $\gamma$ is estimated to be $1.245 \pm 0.008$ for the modified Bethe–type approximation. The fitting for the Weiss–type approximations estimates the value of $\gamma$ to be $1.219 \pm 0.008$, which is inconsistent with the result obtained by the modified Bethe–type approximation. The result from the Weiss–type approximations, however, may also have some deviations, because the results of the two–dimensional case have deviations. The modified Bethe–type approximations will give us the value of the true exponent $\gamma$. So we conclude that the $\gamma$ of the three–dimensional case is $1.245 \pm 0.008$.

When we estimate the critical point and the exponent simultaneously, the critical point is $0.2218 \pm 0.0005$ and the exponent $\gamma$ is $1.26 \pm 0.07$ for the modified Bethe–type approximations.

Also in the three–dimensional case, we can not get meaningful estimations when we estimate the critical point and exponent $\beta$ simultaneously because of the large error associated with $\bar{m}$.

# 5 Summary and Discussion

Our modified Bethe approximation is concluded to be canonical from our results. It is the simplest cluster–effective–field theory. (A simpler theory may be devised but it has almost the same degree–of–approximation.) Its self–consistency can be used in the SEFT and will be helpful in analyzing exotic phase transitions. This approximation is better than the Weiss approximation, because the critical point of a specific cluster in this approximation is closer to the true one than the Weiss one.

The CAM coefficients are calculated for magnetization and susceptibility. We cannot get good estimations for magnetization, but the obtained value of $\beta$ is consistent with the known value. For the susceptibility, good estimations for $\gamma$ and $K_c^*$ can be obtained for the modified Bethe–type approximations. The exponent, $\gamma$, from the Weiss approximations is not consistent with the exact one. Some possible correction terms may be responsible for this discrepancy.

Our estimates for the critical point $K_c$ and exponent $\gamma$ of the three–dimensional Ising ferromagnet are $0.2218 \pm 0.0005$ and $1.26 \pm 0.07$, respectively. If we assume that the critical point is $0.22165$, the estimate of $\gamma$ is $1.245 \pm 0.008$.

It is shown that Monte Carlo simulations can be used to solve the cluster approximations. The combination of Monte Carlo simulations and the effective–field method have also been tried in Refs. [49]. There is an essential difference between our calculations and theirs. We have calculated the quantities defined for finite clusters with free boundary conditions.

Our method will be also useful for some exotic transitions, such as spin glasses (whose physical picture is not clear yet for the finite-dimensional case) and quantum spin systems.

**Acknowledgments**

The authors would like to thank Prof. Y. Kanada for advice on supercomputers. They also thank Dr. M. Katori and Dr. X. Hu for useful discussion. One of them (N. I.) is grateful for financial support of the Fellowships of the Japan Society for the Promotion of Science for Junior Japanese Scientists. He also thanks the Inoue Foundation for Science for financial support to attend this workshop. The simulations were performed on the HITAC S820/80 of the Computer Centre, University of Tokyo and on the HITAC S820/80E of the Hitachi Kanagawa Works. This work is supported by the Research Fund of the Ministry of Education, Science and Culture.

# References

[1] M. Suzuki, J. Phys. Soc. Jpn. **55** (1986) 4205.

[2] M. Suzuki, M. Katori and X. Hu, J. Phys. Soc. Jpn. **56** (1987) 3092.

[3] M. Katori and M. Suzuki, J. Phys. Soc. Jpn. **56** (1987) 3113.

[4] X. Hu, M. Katori and M. Suzuki, J. Phys. Soc. Jpn. **56** (1987) 3865.

[5] X. Hu and M. Suzuki, J. Phys. Soc. Jpn. **57** (1988) 791.

[6] M. Katori and M. Suzuki, J. Phys. Soc. Jpn. **57** (1988) 807.

[7] M. Suzuki, in *Quantum Field Theory*, ed. F. Mancini (Elsevier Science, 1986) 505.

[8] M. Suzuki, Prog. Theor. Phys. Suppl. **87** (1986) 1.

[9] M. Suzuki, Phys. Lett. **A116** (1986) 375.

[10] M. Suzuki, J. Stat. Phys. **49** (1987) 977.

[11] M. Katori and M. Suzuki, in *Progress in Statistical Mechanics*, ed. C. K. Hu (World Scientific, 1988) 273.

[12] M. Suzuki, in *Ordering and Organization in Ionic Solutions*, ed. N. Ise and I. Sogami (World Scientific, 1988) 635.

[13] M. Suzuki, J. Phys. Soc. Jpn. **58** (1989) 3642.

[14] M. Suzuki, Phys. Lett. **A127** (1988) 410.

[15] M. Suzuki, J. Phys. Soc. Jpn. **56** (1987) 4221.

[16] M. Suzuki, J. Phys. Soc. Jpn. **57** (1988) 1.

[17] M. Katori and M. Suzuki, J. Phys. Soc. Jpn. **57** (1988) 3753.

[18] J. L. Monroe, Phys. Lett. **A131** (1988) 427.

[19] T. Oguchi and H. Kitatani, J. Phys. Soc. Jpn. **58** (1989) 3033.

[20] N. Ito and M. Suzuki, Int. J. Mod. Phys. **B2** (1988) 1.

[21] T. Oguchi and H. Kitatani, J. Phys. Soc. Jpn. **57** (1988) 3973.

[22] M. Katori, J. Phys. Soc. Jpn. **57** (1988) 4114.

[23] S. Fujiki, in *Proc. 2nd. YKIS on Cooperative Dynamics in Complex Physical Systems*, ed. H.Takayama (Springer, 1988) 179.

[24] X. Hu and M. Suzuki, Physica **A150** (1988) 310.

[25] M. Takayasu and H. Takayasu, Phys. Lett. **A128** (1988) 45.

[26] H. Takayasu, M. Takayasu and T. Nakamura, Phys. Lett. **A132** (1988) 429.

[27] N. Konno and M. Katori, J. Phys. Soc. Jpn. **59** (1990) 1581.

[28] N. Ito and Y. Kanada, Supercomputer **5** No.3 (1988) 31.

[29] N. Ito and Y. Kanada, Supercomputer **7** No.1 (1990) 29.

[30] M. Suzuki, J. Phys. Soc. Jpn. **57** (1988) 683.

[31] M. Suzuki, J. Phys. Soc. Jpn. **57** (1988) 2310.

[32] M. Suzuki, J. Stat. Phys. **53** (1988) 483.

[33] M. Suzuki, in *Dynamics of Ordering Processes in Condensed Matter*, ed. S. Komura and H. Furukawa (Plenum, 1988) 23.

[34] M. Suzuki, in *Proc. 2nd. YKIS on Cooperative Dynamics in Complex Physical Systems*, ed. H.Takayama (Springer, 1988) 9.

[35] M. Suzuki, J. de Physique, Colloque C8, Supplement. Tome 49 (1988) 1591.

[36] N. Kawashima and M. Suzuki, J. Phys. Soc. Jpn.**58** (1989) 3123.

[37] N. Ito, M. Taiji and M. Suzuki, J. Phys. Soc. Jpn. **56** (1987) 4218..

[38] N. Ito, M. Taiji and M. Suzuki, J. de Physique, Colloque C8, Supplement. Tome 49 (1988) 1397.

[39] R. B. Pearson, J. L. Richardson and D. Toussaint, Phys. Rev. **31** (1985) 4472.

[40] B. M. McCoy and T. T. Wu, *The Two-dimensional Ising Model*, (Harvard Univ. Press, 1973).

[41] G. A. Baker, Jr., Phys. Rev. **124** (1961) 768.

[42] J. W. Essam and M. E. Fisher, J. Chem. Phys. **38** (1963) 802.

[43] G. S. Pawley, R. H. Swendsen, D. J. Wallace and K. G. Wilson, Phys. Rev. **B29** (1984) 4030.

[44] M. N. Barber, P. B. Pearson, D. Toussaint and J. L. Richardson, Phys. Rev. **B32** (1985) 1770.

[45] D. S. Gaunt and M. F. Sykes, J. Phys. **A12** (1979) L25.

[46] J. Zinn–Justin, J. Physique **40** (1979) 969.

[47] F. Lee and H. H. Chen, Phys. Lett. **82A** (1981) 140.

[48] J. C. LeGuillou and J. Zinn–Justin, J. Physique Lett. **46** (1985) L137.

[49] H. Müller–Krumbhaar and K. Binder, Z. Phys. **254** (1972) 269.

*Note added in proof*– The correct behaviors of coherent anomalies for cluster–type approximations are discovered. The details are given in the paper: N. Ito and M. Suzuki, to be published in Phys. Rev. B. The Weiss–type approximation can also produce correct exponents if the correct formulation is applied for analysis.

# Histogram Techniques for Studying Phase Transitions

*A.M. Ferrenberg*

Center for Simulational Physics, The University of Georgia,
Athens, GA 30602, USA

Abstract: The use of histogram techniques to increase the efficiency of Monte Carlo computer simulations is discussed. The early history, as well as more recent refinements are presented. The power of the technique is demonstrated by Monte Carlo studies of three statistical mechanical models.

1. Introduction

As the Monte Carlo (MC) simulation method [1] has grown in popularity, it has been used to study systems of increasing complexity [2–3]. At the same time, the standards for accuracy in MC studies have been raised as the method has matured. In order to meet the computational demands imposed by these two considerations, a great deal of effort has gone into increasing the efficiency of MC simulations.

There are two main approaches to this problem. The first is to increase the amount of data which can be generated in a given amount of computer time, or to reduce the effects of critical slowing down near phase transitions. Advances in computer hardware, for example the development of vector and parallel computers, and simulation algorithms, such as multi–spin coding [4] and cluster–flipping algorithms [5], have made it possible to study larger systems with higher accuracy than ever before.

A complementary approach to increasing the efficiency is to extract more information from the data generated during the simulation. A well–known example of this kind of approach is the Monte Carlo Renormalization Group (MCRG) method [6] which uses MC data to directly calculate critical exponents.

In this paper, we would like to describe **histogram techniques** which follow this second approach by using the data generated in a MC simulation at one temperature to determine the properties of the system over a range of temperatures. Section 2 gives the theory behind histogram techniques including a brief history of the use of histograms in MC simulations. Two technical aspects of the implementation of the method are discussed in

Section 3. The power and general applicability of histogram techniques are demonstrated in Section 4 where some results of MC studies of three lattice models— a d=2 dimer model, the d=3 classical Heisenberg model and the d=3 Ising model— are presented. Section 5 contains concluding remarks.

## 2. Histogram Techniques

The histogram technique presented here has been described elsewhere [7]. In this Section, we will briefly outline the method and introduce the notation needed later.

A Monte Carlo simulation is a numerical method for generating configurations of a physical system based on a certain probability distribution. In the following, we will be concerned with constant temperature MC simulations. Such a simulation, performed at temperature $T = T_0$, generates configurations according to the canonical distribution

$$P_{\beta_0}(X) = \frac{1}{Z(\beta_0)} \exp[-\beta_0 \mathcal{H}(X)] \;, \qquad (1)$$

where X represents a system configuration, $\beta_0 = 1/k_B T_0$, ($k_B$ is Boltzmann's constant) $\mathcal{H}(X)$ is the Hamiltonian of the system being studied, and the normalizing factor $Z(\beta_0) = \sum_{\{X\}} \exp[-\beta_0 \mathcal{H}(X)]$ is the partition function.

Although the probability distribution $P_{\beta_0}(X)$ contains the complete thermodynamic information about the system, it is more convenient to work with a related distribution, $P_{\beta_0}(E)$, where E is one of the values in the spectrum (continuous or discrete) of $\mathcal{H}(X)$:

$$P_{\beta_0}(E) = \frac{1}{Z(\beta_0)} W(E) \exp[-\beta_0 E] \;. \qquad (2)$$

The number (density) of states,

$$W(E) = \sum_{\{X\}} \delta_{E, \mathcal{H}(X)} \;,$$

where $\delta$ represents the Kronecker delta function, depends only on the structure of the phase space of the model and is therefore independent of $\beta_0$. For a system with a continuous energy spectrum, the sum over states is replaced by an integral. In practice, however, one studies the distribution by binning it, choosing the number of bins (and their size if non–uniform bins are used) to minimize discretization effects. If the Hamiltonian consists of more than one term, $\mathcal{H}(X) = \mathcal{H}_1(X) + \mathcal{H}_2(X) + \cdots$, the density of states depends on more than one variable, that is $W = W(E_1, E_2, \cdots)$. To simplify the notation, we will consider only a single–term Hamiltonian in this discussion but will return to the case of a multi–term Hamiltonian in Section 3.

The average value of any function of E can be calculated from (2):

$$<f(E)>_{\beta_0} = \frac{1}{Z(\beta_0)} \sum_E f(E) \, W(E) \exp[-\beta_0 E] \,. \tag{3}$$

Because the simulation generates configurations according to the equilibrium probability distribution, an energy histogram, H(E), kept during the simulation provides an estimate for the equilibrium distribution (2), becoming exact in the limit of an infinite run. For a finite simulation, the histogram will be subject to statistical errors, but H(E)/N, where N is the number of measurements made, still provides an estimate for (2) so that we can write

$$H(E) = \frac{N}{Z(\beta_0)} W(E) \exp[-\beta_0 E] \,. \tag{4}$$

From the form of (2) and (4), it is clear that knowledge of the distribution at one value of $\beta$ is sufficient, in principle, to determine it for any value of $\beta$. To see this, recall that the probability distribution for arbitrary $\beta$ has the same form as (2):

$$P_\beta(E) = \frac{1}{Z(\beta)} W(E) \exp[-\beta E] \,. \tag{5}$$

Using the histogram from the simulation performed at $\beta_0$, we can estimate W(E) by inverting (4)

$$W(E) = \frac{Z(\beta_0)}{N} H(E) \exp[\beta_0 E] \,. \tag{6}$$

By inserting this estimate for W(E) into (5), and forcing the distribution to be normalized, the relationship between the histogram measured at $\beta = \beta_0$ and the probability distribution for arbitrary $\beta$ becomes

$$P_\beta(E) = \frac{H(E) \exp[-\Delta\beta \, E]}{\sum_E H(E) \exp[-\Delta\beta \, E]} \,, \tag{7}$$

where $\Delta\beta = (\beta - \beta_0)$. The average of any function of E (3) is then

$$<f(E)>_\beta = \frac{\sum_E f(E) \, H(E) \exp[-\Delta\beta \, E]}{\sum_E H(E) \exp[-\Delta\beta \, E]} \,. \tag{8}$$

Equations (7) and (8) will be referred to as the *single–histogram equations*.

Because of the finite length of MC simulations, the single–histogram equations provide reliable results only for a finite $\Delta\beta$ around the simulated value $\beta_0$. When $\Delta\beta$ becomes too large in magnitude, the statistical inaccuracy in the wings of the histogram becomes sufficiently important that the method may yield unreliable results. By performing a small number of additional simulations, however, one can guarantee that the results obtained using (7) and (8) do not suffer from systematic errors.

The idea of using the histogram generated in a MC simulation is not new. The first application dates back to 1959 when Salsburg and co-workers [8] incorporated these ideas in a MC study. They wrote down the necessary equations and explained that a histogram of values of an observable from a MC simulation could be used to evaluate any function of that observable at any neighboring value of the corresponding parameter. Curiously enough, they evaluated the equation only at the temperature of the original simulation which provided no additional information. Four years later, Chesnut and Salsburg [9] described the use of histograms to obtain information over a range of continuously varying parameters, but they also failed to implement this idea.

In 1967, McDonald and Singer [10] were the first to actually use the single-histogram equations in a MC study. They simulated a system of liquid argon with a Lennard-Jones 12-6 potential. Using the single-histogram equations, which they called "variation of Lennard-Jones parameters", along with certain approximations, they were able to extrapolate data taken for (V,T) (V being the volume of the system) to (V',T) by "isothermal reweighting" and (V,T) to (V,T') by "isochoric reweighting".

Following this, interest in single-histogram techniques waned while much effort went into the development of multiple-histogram methods. A brief review of this work can be found in [11].

Interest in single-histogram techniques was renewed in 1982 and 1984 in work performed by Falcioni and co-workers [12] and Marinari [13] who used the single-histogram method of McDonald and Singer but calculated properties of the system for complex temperature values. In particular, they measured the location of zeroes of the partition function to estimate the critical temperature as well as critical exponents and amplitude ratios.

In 1988, Ferrenberg and Swendsen [7] showed that histograms could be used to efficiently determine the thermodynamic properties of systems, particularly near a phase transition where the behavior of the system exhibits narrow peaks in thermodynamic derivatives. Standard MC techniques, which provide data only for the simulated temperature, are inefficient for locating these sharp peaks with high accuracy. Ferrenberg and Swendsen demonstrated that the data from a single simulation in the vicinity of a peak could be used to locate the peak with extremely high accuracy.

This new understanding of the single–histogram method, as well as the development of several multiple histogram methods [11,14–20], has led to the increased use of histogram techniques in the MC study of statistical models [21].

## 3. Implementation of the Single-Histogram Method

Although the single–histogram equations (7) and (8) are quite simple, certain practical problems can arise in their implementation. In this section, we discuss two of these and provide algorithms to overcome them.

The first problem is the evaluation of the exponential $\exp[-\Delta\beta\,E]$. Because the energy E is an extensive quantity, proportional to the volume of the system, one is faced with the possibility that the argument of the exponential $[-\Delta\beta\,E]$ will become so large that its evaluation leads to floating–point overflow. Fortunately, the solution to this problem is straightforward.

As a preliminary remark, we note that it is convenient, conceptually as well as algorithmically, to work with the logarithm of the histogram rather than the histogram itself. Introducing the notation $Q(E) = \log H(E)$, the expression for the average value of a function of E (8) becomes

$$<f(E)>_\beta = \frac{\sum_E f(E)\,\exp[Q(E) - \Delta\beta\,E]}{\sum_E \exp[Q(E) - \Delta\beta\,E]}. \quad (9)$$

For each value of $\beta$, we determine the maximum value of the argument of the exponential

$$\mathrm{ArgMax} = \max_E (Q(E) - \Delta\beta\,E).$$

Equation (9) can now be re–written as

$$<f(E)>_\beta = \frac{\exp[\mathrm{ArgMax}]\sum_E f(E)\,\exp[Q(E) - \Delta\beta\,E - \mathrm{ArgMax}]}{\exp[\mathrm{ArgMax}]\sum_E \exp[Q(E) - \Delta\beta\,E - \mathrm{ArgMax}]}$$

or, canceling the term exp[ArgMax] from the numerator and denominator,

$$<f(E)>_\beta = \frac{\sum_E f(E)\,\exp[Q(E) - \Delta\beta\,E - \mathrm{ArgMax}]}{\sum_E \exp[Q(E) - \Delta\beta\,E - \mathrm{ArgMax}]}. \quad (10)$$

The maximum value of the argument of the exponential is now zero so overflow is eliminated. The expression can be checked for underflow making the calculation numerically stable. In most situations, Q(E) is quite small ($\leq 15$) so that any overflow is due to the term ($\Delta\beta$ E) and ArgMax can be determined more simply by

$$\text{ArgMax} = \begin{cases} -\Delta\beta\, E_{min} & \text{for } \Delta\beta > 0 \\ -\Delta\beta\, E_{max} & \text{for } \Delta\beta < 0 \end{cases}$$

where $E_{min}$ and $E_{max}$ are the smallest and largest energies generated during the simulation.

The second problem, which arises for large systems with multi–parameter Hamiltonians, is that of computer memory. Because each term in the Hamiltonian is, in general, extensive, the amount of memory needed to store the entire histogram for an n–term Hamiltonian increases with the volume like $V^n$. This can easily exceed the available memory on even the largest computers for moderate sized systems. This problem can also be overcome relatively easily, provided the simulation data are stored properly. The most simple solution is to have the simulation program write out the energy terms $E_1$, $E_2$ $\cdots$ to a file during the simulation rather than trying to store the histogram. The necessary histograms can later be constructed from this data as shown below. In addition to this, data stored in this manner can also be used for correlation time measurements, a necessary part of any careful MC study. The data can also be binned to study the statistical and systematic errors which arise due to the finite length of the MC runs [22]. Although storing the data in this manner can require large amounts of disk or tape storage, the benefits of keeping the data are great, and efficient data compression algorithms can be used to reduce the amount of storage needed.

The technique for overcoming the problem of computer memory is best illustrated by a concrete example. Consider the Hamiltonian for the d = 3 nearest neighbor Ising model

$$\mathcal{H} = -J \sum_{n.n.} \sigma_i \sigma_j - B \sum_i \sigma_i \quad , \tag{11}$$

where $\sum_{n.n.}$ denotes the sum over all nearest–neighbor pairs, $\sum_i$ is a sum over all lattice sites, J is the coupling constant and B is an applied magnetic field. The Hamiltonian is more commonly expressed in terms of the dimensionless coupling constant $K = \beta J$ and dimensionless field $h = \beta B$, that is

$$-\beta \mathcal{H} = K \sum_{n.n.} \sigma_i \sigma_j + h \sum_i \sigma_i = K\,E + h\,M \tag{12}$$

with $E = \sum_{n.n.} \sigma_i \sigma_j$ and $M = \sum_i \sigma_i$. With the data stored in the two–dimensional histogram H(E,M) it would be possible to calculate the properties of the system for a range of K and h around the simulated values $K_0$ and $h_0$. This could be done, for example, by fixing the magnetic field to some value h and then varying K. By repeating the process for different values of h one is able to scan an entire $(\Delta K, \Delta h)$ region.

Measured quantities of interest can usually be expressed as the product of a function of E and a function of M, $g(E)f(M)$. For example, the derivative with respect to K of any function of M can be calculated as

$$\frac{\partial <f(M)>_K}{\partial K} = <E\,f(M)>_K - <E>_K <f(M)>_K \,. \tag{13}$$

The average value of $g(E)f(M)$ is determined by

$$<g(E)f(M)>_{K,h} = \frac{\sum\limits_{E,M} g(E)f(M)H(E,M)\exp[\Delta K\,E + \Delta h\,M]}{\sum\limits_{E,M} H(E,M)\exp[\Delta K\,E + \Delta h\,M]} \,, \tag{14}$$

which, by performing the sums over M first, can be re-expressed as

$$<g(E)f(M)>_K = \frac{\sum\limits_{E} g(E) <f>_h(E)\,\tilde{H}(E)\exp[\Delta K\,E]}{\sum\limits_{E} \tilde{H}(E)\exp[\Delta K\,E]} \,, \tag{15}$$

where $\tilde{H}(E) = \sum\limits_{M} H(E,M)$ is a one-dimensional histogram which should easily fit into the computer memory and can be calculated from the list of data stored from the simulation. The quantity $<f>_h(E)$ is the microcanonical average of $f(M)$ estimated from the simulation data. This is calculated from the list of data in the following manner. First, let $E_t$ and $M_t$ designate the values of E and M at MC time t. Then

$$<f>_h(E) = \frac{\sum\limits_{t} f(M_t)\exp[\Delta h\,M_t]\,\delta_{E,E_t}}{\sum\limits_{t} \delta_{E,E_t}}$$

$$= \frac{\sum\limits_{t} f(M_t)\exp[\Delta h\,M_t]\,\delta_{E,E_t}}{\tilde{H}(E)} \,. \tag{16}$$

A separate average $<f>_h(E)$ is calculated for each function of M needed, for example M, $|M|$, $M^2$ and $M^4$. Using this method, the magnetic properties of a 96×96×96 Ising lattice have been calculated as a function of K with a program running on a personal computer

[23]. (The simulations themselves were previously performed on a Cyber 205 vector computer [24].) Some results of this study will be presented in the next section.

## 4. Applications of the Single-Histogram Method

In this section, we present a sampling of the results from three MC studies carried out primarily at the University of Georgia. The results will be presented in detail elsewhere.

The first model we consider is a two-dimensional lattice dimer model proposed by Nagle and co-workers [25]. The system consists of dimers on a square lattice, with no vacancies, and a long-range interaction with the Hamiltonian

$$\mathcal{H} = -(b/N)(N_h - N_v)^2,$$

where $N_h$ and $N_v$ are, respectively, the number of horizontal and vertical dimers, b is the strength of the interaction, and $N = N_v + N_h$ is the total number of dimers. Because N is constant, the Hamiltonian can be expressed in terms of a single variable $N_v$:

$$\mathcal{H} = -(b/N)(2N_v - N)^2 = -Nb\rho^2$$

where the polarization $\rho = (2N_v - N)/N$. Although this system is quite simple, and has been solved exactly [25], we chose to simulate it [26] because it possesses some of the important properties of more realistic dimer and polymer models. The most important property is the presence of "hard" (excluded volume) interactions in this model. Unlike spin systems, where the energy cost for flipping a spin is of the order of $k_B T_c$ near the phase transition, dense polymer systems contain interactions, arising when two particles attempt to occupy the same volume, that are very large compared to $k_B T_c$. In this dimer model, we assign an infinite energy to configurations in which two dimers attempt to occupy the same lattice site. While this makes the model more realistic, it becomes extremely difficult to simulate because any changes in the system configuration must involve at least two dimers. (Moving any single dimer results in overlap, because the model contains no vacancies, and therefore an infinite energy.) The non-local nature of the algorithm needed to overcome this difficulty [27] makes the simulation quite slow so that efficiency in the analysis of the data is crucial.

Using the data from simulations performed at the infinite-lattice transition temperature, $k_B T_c/b = 8/\pi$, we can determine the location of the maximum of the specific heat, as well as the location of the maximum slope in the energy cumulant $V_L$ and the polarization cumulant $U_L$. The fourth-order cumulant is defined [28] by

$$U_L = 1 - \frac{<\rho^4>}{3<\rho^2>^2} \quad . \tag{17}$$

The energy cumulant is calculated in the same manner, but with $\rho$ replaced by the energy $E = -bN\rho^2$ in (17). The temperatures where these three maxima occur can be used as estimates for the finite–lattice transition temperature $T_c(L)$. From finite–size scaling theory [29] we expect that $T_c(L)$ will approach $T_c$ like

$$T_c(L) = T_c + aL^{-1/\nu} \quad . \tag{18}$$

The value of $a$ depends on the particular quantity used to estimate $T_c(L)$ and $\nu$ is the exponent describing the divergence of the correlation length in the infinite system. These three estimates for $T_c(L)$ are plotted vs. $L^{-1/\nu}$ in Figure 1. The exact infinite lattice transition temperature is marked by an arrow. The exponent $\nu$ was calculated by studying the finite–size scaling of the maximum slope of $U_L$ to be $\nu = 0.99 \pm 0.01$ in excellent agreement with the exact result $\nu = 1$.

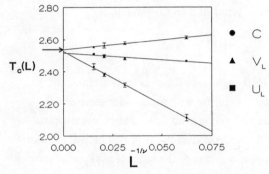

Fig. 1    Estimates for $T_c(L)$ from the specific heat maximum and the location of the maximum slopes of the polarization cumulant $U_L$ and energy cumulant $V_L$ for the d = 2 dimer model. The exact infinite lattice transition temperature is marked with an arrow.

The second model considered here is the (classical) d = 3 Heisenberg model with the Hamiltonian

$$\mathcal{H} = -J \sum_{n.n.} \vec{S}_i \cdot \vec{S}_j \quad .$$

J is the coupling constant and the sum goes over all nearest neighbor pairs. This model is provides an excellent test for histogram techniques because its energy spectrum is continuous so the histogram must be binned. The careful study of the static critical properties of this model using histogram techniques [30] was encouraged by the results of a study of the dynamic critical behavior [31] in which estimates for the static exponent ratios

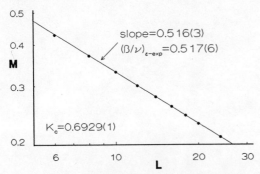

Fig. 2   Magnetization vs. system size for the d = 3 Heisenberg model at the new estimate for the inverse critical temperature 0.6929. The value of $\beta/\nu = 0.516 \pm 0.003$ obtained from the slope is in excellent agreement with $\epsilon$-expansion results [32].

$\beta/\nu$ and $\gamma/\nu$ were inconsistent with previous $\epsilon$-expansion results [32] ($\beta/\nu$ was determined to be $0.550 \pm 0.006$ which differs substantially from the result $0.517 \pm 0.006$ [32]). A possible explanation for the discrepancy was an incorrect value for $T_c$. To test this possibility, the simulation data, which had been stored in the manner described in Section 3, were re-analyzed using single-histogram [7] and multiple histogram [18] methods. From careful analysis, an estimate for the inverse critical temperature $J/k_B T_c$ of $0.6929 \pm 0.0001$ was obtained. This result differs considerably from the value 0.6916 obtained from high-temperature series expansions [33] used for the original simulations, but agrees better with more recent transfer-matrix MC results of $0.6922 \pm 0.0002$ and $0.6925 \pm 0.0003$ [34]. In Figure 2, the magnetization calculated at the new estimate for $T_c$ is plotted vs. system size L to extract $\beta/\nu$. The result $\beta/\nu = 0.516 \pm 0.003$ is in excellent agreement with the $\epsilon$-expansion result [32] as are estimates for other critical exponents [30].

The final example we consider here is the d = 3 Ising model with the Hamiltonian (12). Despite intense effort, the model has defied solution even for h = 0. Numerical analyses, using series expansion [35–36] and MCRG [37–39] techniques have provided impressive results for the critical coupling $K_c$, although no definitive value has yet been determined.

Monte Carlo simulations using finite-size scaling analyses [40–43] have not, up to now, been able to compete with other numerical techniques in providing accurate estimates for $K_c$. Histogram techniques, as presented in this paper, have the potential to alter this situation.

We [23] have recently begun a histogram analysis of simulation data taken for the d = 3 Ising model several years ago for a study of critical dynamics [44]. (We were fortunate that these simulation results had been stored in the manner described in Section 3.) Although

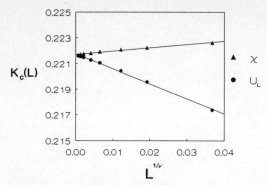

Fig. 3    Estimates for $K_c(L)$ for the d = 3 Ising model estimated from the maximum values of the cumulant slope and finite–lattice susceptibility.

our analysis of this data is incomplete, we have preliminary estimates for $K_c(L)$, the finite–lattice critical coupling, from two sources: the location of the maximum slope of the 4th order magnetization cumulant $U_L$ and the maximum of the finite–lattice susceptibility $\chi = KL^3[<m^2> - <|m|>^2]$. Results for $K_c(L)$ from these two quantities are shown in Figure 3. The value of $\nu$ used was determined from the size dependence of the cumulant slope, and is in excellent agreement with MCRG results [37–39]. Our early results are promising, although more careful analysis is needed before a final answer can be obtained.

5. Conclusions

We have presented an overview of the use of histogram techniques in the study of critical phenomena. Although these methods have been known for more than thirty years, they have only recently been applied with success to the study of critical phenomena. It is hoped that the current interest in these methods will lead to further refinements and the development of even more efficient uses of simulation data.

The author wishes to acknowledge fruitful collaboration with C. Schiff, J.G. Fetkovich, J.F. Nagle, P. Peczak and D.P. Landau on the work presented in Section 4. This work was supported, in part, by NSF grant DMR–8715740.

REFERENCES

1. N. Metropolis, A.W. Rosenbluth, M.N. Rosenbluth, A.H. Teller and E. Teller, J. Chem. Phys. 21, 1087 (1953).
2. "Monte Carlo Methods in Statistical Physics", Second Edition, K. Binder, ed., (Springer–Verlag, New York, 1986).

3. K. Binder, J. Comp. Phys. **59**, 1 (1985).
4. A discussion of these topics can be found in: K. Binder, "Recent Trends in the Development and Application of the Monte Carlo Method", in Reference 2.
5. R.H. Swendsen and J.–S. Wang, Phys. Rev. Lett. **58**, 86 (1987).
6. R.H. Swendsen, Phys. Rev. Lett. **42**, 859 (1979).
7. A.M. Ferrenberg and R.H. Swendsen, Phys. Rev. Lett., **61**, 2635 (1988).
8. Z.W. Salsburg, J.D. Jackson, W. Fickett and W.W. Wood, J. Chem. Phys. **30**, 65 (1959).
9. D.A. Chesnut and Z.W. Salsburg, J. Chem. Phys. **38**, 2861 (1963).
10. R. McDonald and K. Singer, Disc. Far. Soc, **43**, 400 (1967).
11. A.M. Ferrenberg and R.H. Swendsen, "Optimized Monte Carlo Data Analysis", Computers in Physics, Vol. 3, No. 5, (1989).
12. M. Falcioni, E. Marinari, M.L. Paciello, G. Parisi and B. Taglienti, Phys. Lett. **108B**, 331 (1982).
13. E. Marinari, Nucl. Phys. **B235[FS11]**, 1233 (1984).
14. G. Bhanot, S. Black, P. Carter and R. Salvador, Phys. Lett. **B183**, 331 (1987).
15. G. Bhanot, K.M. Bitar, S. Black, P. Carter and R. Salvador, Phys. Lett. **B187**, 381 (1987).
16. G. Bhanot, K.M. Bitar and R. Salvador, Phys. Lett. **B188**, 246 (1987).
17. G. Bhanot, R. Salvador, S. Black and R. Toral, Phys. Rev. Lett. **59**, 803 (1987).
18. A.M. Ferrenberg and R.H. Swendsen, Phys. Rev. Lett., **63**, 1195 (1989).
19. P.B. Bowen, J.L. Burke, P.G. Corsten, K.J. Crowell, K.L. Farrell, J.C. Macdonald, R.P. Macdonald, A.B. MacIsaac, S.C. MacIsaac, P.H. Poole and N. Jan, Phys. Rev. B **40**, 3454 (1989).
20. N.A. Alves, B.A. Berg and R. Villanova, Phys. Rev. B **41**, 383 (1990).
21. A review of past and present applications of histograms can be found in: A.M. Ferrenberg and R.H. Swendsen, "Histogram Methods in Statistical Physics", submitted to Int. Rev. Mod. Phys.
22. A.M. Ferrenberg, D.P. Landau and K. Binder, submitted to J. Stat. Phys.
23. A.M. Ferrenberg and D.P. Landau, preprint.
24. S. Wansleben, Comp. Phys. Comm. **43**, 315 (1987).
25. J.F. Nagle, A. Yanagawa and J. Stecki, Mol. Cryst. Liq. Cryst, **37**, 127 (1976).
26. A.M. Ferrenberg, C. Schiff, J.G. Fetkovich and J.F. Nagle, preprint.
27. The algorithm used is a variation of one proposed in: A. Rahman and F.H. Stillinger, J. Chem. Phys. **57**, 4009 (1972).
28. K. Binder, Z. Phys. B **43**, 119 (1981).
29. M.E. Fisher, Proceedings of the International Summer School "Enrico Fermi" Course 51 (Academic, New York, 1971).
30. P. Peczak, A.M. Ferrenberg and D.P. Landau, preprint.

31. P. Peczak and D.P. Landau, J. Appl. Phys. **67**, 5427 (1990).
32. J.C. LeGuillou and J. Zinn–Justin, Phys. Rev. B **21**, 3976 (1980); J. Phys. Lett. **46**, L137 (1985).
33. D.S. Ritchie and M.E. Fisher, Phys. Rev. B **5**, 2668 (1972).
34. M.P. Nightingale and H.W.J. Blöte, Phys. Rev. Lett. **60**, 1562 (1988).
35. B.G. Nickel, in Phase Transitions: Cargese 1980, M. Levy, J.–C. LeGuillou and J. Zinn–Justin, eds. (Plenum, New York, 1982).
36. A.J. Liu and M.E. Fisher, Physica A **165**, 35 (1989).
37. G.S. Pawley, R.H. Swendsen, D.J. Wallace and K.G. Wilson, Phys. Rev. B **29**, 4030 (1984).
38. H.W.J. Blöte, J. de Bruin, A. Compagner, J.H. Croockewit, Y.T.J.C. Fonk, J.R. Heringa, A. Hoogland, and A.L. van Willigen, Europhys. Lett. **10**, 105 (1989).
39. H.W.J. Blöte, A. Compagner, J.H. Croockewit, Y.T.J.C. Fonk, J.R. Heringa, T.S. Smit, and A.L. van Willigen, Physica **A161**, 1 (1989).
40. D.P. Landau, Phys. Rev. B **14**, 255 (1976).
41. M.N. Barber, R.B. Pearson, D. Toussaint, and J.L. Richardson, Phys. Rev. B **32**, 1720 (1985).
42. G. Parisi and F. Rapuano, Phys. Lett. **B157**, 301 (1985).
43. G. Bhanot, D. Duke, and R. Salvador, Phys. Rev. B **33**, 7841 (1986).
44. S. Wansleben and D.P. Landau, J. Appl. Phys. **61**, 3968 (1987) and unpublished.

# Classification of Cellular Automata

*R.W. Gerling*

Institut für Theoretische Physik I der Universität Erlangen-Nürnberg,
Staudtstr. 7, W-8520 Erlangen, Fed. Rep. of Germany and
HLRZ, c/o KFA Jülich, Postfach 1913, W-5170 Jülich, Fed. Rep. of Germany

**Abstract.** Cellular automata are studied on a variety of different lattices. The results for the classification and for damage spreading are reported. Details of the algorithm and the implementation on different computers are also presented.

## 1. Introduction

Cellular automata are of quite some recent interest, ranging in their applications from hydrodynamics [1] to biological systems [2]. In a cellular automaton each site of a given lattice carries a variable $\sigma_i(t)$, which can have one of two values namely zero or one. The new value of the variable at time $t+1$ is given entirely by the value of the nearest neighbors. The table which gives the new value $\sigma_i(t+1)$ for all possible configurations of the neighbors is called the rule of the cellular automaton. On a lattice with $z$ nearest neighbors we have $m = 2^z$ possible neighborhoods and therefore $2^m$ possible rules.

The rules for one-dimensional and the totalistic two-dimensional square lattice cellular automata were classified by Wolfram [3,4] depending upon the final state of the application of the rule. Stauffer [5] did the first systematic study of two-dimensional cellular automata. Manna and Stauffer [6] extended the Wolfram classification scheme to seven classes. Their classes are:

Class 0:  Fixed point with all spins down.
Class 1:  Fixed point with all spins up.
Class 2:  Fixed point with some spins up and some spins down.
Class 3:  Oscillations between the states with all spins up and all spins down.
Class 4:  Oscillations between two different states with some spins up and some spins down.
Class 5:  Moving structures.
Class 6:  All rules which show other behavior than classes 0 to 5.

We have studied the one-dimensional lattice, the two-dimensional honeycomb and triangular lattices and the three-dimensional simple cubic lattice [7,8].

The other aspect of our study is damage spreading. Two lattices are both updated simultaneously with the same rule. The initial configurations of the two lattices were identical except a few spins in the center of the lattice, which were flipped. The sites where the two lattices differ in the later simulation are called the damage. We studied the time development of the damage.

Stauffer [5] classified the damage spreading into three groups. From the results of our studies [7,8] Stauffer's class $b$ was further subdivided:

Class $a$: The damage heals out after some time, and both lattices become the same. Possible candidates for this class are the rules which lead to fixed points in the other classification. We call rules, which lead to this class, stable.

Class $b_0$: The damage does not heal out, but its size remains small. The damage is localized and does not move.

Class $b_c$: The damage does not heal out, but the damage moves over the whole lattice. It remains small.

Class $c$: The damage spreads over the whole lattice. A rule, which leads to this class, is called unstable.

## 2. Implementation

We used a fully vectorized variant of the da Silva and Herrmann [9] multi-site coding algorithm to update the lattice. One bit is used for one site and 64 bits in one Cray word are updated in parallel. A lookup table is used to relate a given bit in the rule to the corresponding configuration of the neighborhood. The various lattices are implemented with periodic boundary conditions in all directions.

The programs were tested with simple rules, which could be worked out by hand. A simple example is the bureaucratic rule, where the central site depends only on the neighbor on top and the other neighbors are ignored. This rule leads to a simple shift of the whole lattice.

The one-dimensional cellular automata program for a scalar computer is given in Listing 1. This program is very well suited to be run on a plain PC for instructional purposes (Set `ibit=32` in line 2 and adjust `mdim` accordingly for an IBM PC). Listing 2 shows the same program modified to be run on a vector computer. Using the CFT77 compiler on a Cray Y-MP the program fully vectorizes. The reader can easily identify the rearrangements in the code to vectorize the program.

In Listing 3 a one-dimensional cellular automata program for the Connection Machine is shown. This program was used to study the suitability of the CM-2 parallel computer to study cellular automata. The syntax is the array-syntax of the draft ANSI standard Fortran 8x.

For coordination numbers up to 4 simply all rules were simulated. For the two lattices with coordination number 6 a systematic study of all rules is not possible. In the program 1000 randomly chosen rules are applied to a random starting configuration of spins, which are with probability $p$ up and with probability $1 - p$ down. For all simulations we used $p = 0.5$. The same starting configuration was used for all rules. As long as $p$ is not changed the influence of the starting configuration can be neglected. Going to a different set

**Listing 1:** *An implementation of the da Silva-Herrmann algorithm for a scalar computer. Any extensions to Fortran 77 are in the syntax of the Cray Fortran CFT77.*

```
  1  C       systematic investigation of 1-dim cellular automata
  2          PARAMETER (mdim=32000,ibit=64,mdimp1=mdim+1,ldim=ibit*mdim)
  3          DIMENSION nr(0:3),n(0:mdimp1),nf(mdim),nstart(mdim)
  4          DATA mt,iseed,pstart/200,13579,0.5/
  5          CALL RANSET(2*iseed-1)
  6          DO 1 j=1,mdim
  7            nstart(j)=0
  8            DO 1 ii=1,ibit
  9  1         nstart(j)=SHIFT(nstart(j).OR.IFIX(pstart+1.0-RANF()),1)
 10          DO 2 ica=0,15
 11            DO 3 k=0,3
 12  3         nr(k)=1.AND.SHIFT(ica,ibit-k)
 13            DO 4 j=1,mdim
 14  4         n(j)=nstart(j)
 15  c====== end of initialization, start of time development =======
 16            DO 5 it=1,mt
 17              n(0)=SHIFT(n(mdim),ibit-1)
 18              n(mdimp1)=SHIFT(n(1),1)
 19              DO 6 j=1,mdim
 20                nf(j)=0
 21                IF(nr(0).EQ.1) nf(j)=nf(j).OR.(.NOT.n(j+1).AND..NOT.n(j-1))
 22                IF(nr(1).EQ.1) nf(j)=nf(j).OR.(     n(j+1).AND..NOT.n(j-1))
 23                IF(nr(2).EQ.1) nf(j)=nf(j).OR.(.NOT.n(j+1).AND.     n(j-1))
 24  6           IF(nr(3).EQ.1) nf(j)=nf(j).OR.(     n(j+1).AND.     n(j-1))
 25  c======== end of simulation, start updating and analysis =======
 26              m=0
 27              DO 7 j=1,mdim
 28                n(j)=nf(j)
 29  7           m=m.OR.nf(j)
 30  5         IF (m.EQ.0) GOTO 2
 31  2      WRITE (6,*) ica,m,mt,ldim,iseed,pstart
 32        END
```

of random rules showed some fluctuation in the results. By averaging the results over 10 runs with 1000 rules each we got averages and estimates for the errors.

We also measured the performance of different computers with the one-dimensional cellular automata program. On a PC (12 MHz 80286) we reached about 250 spin-updates (SU)/millisecond. One processor of the Cray Y-MP performed at a speed of 985 SU/microsecond for a lattice with 2048000 spins. The CM-2 ran at about 338 SU/microsecond for a lattice of $2^{21}$ spins on 16382 processors. The fact that a CM-2 with 16382 processors is about one third of the speed of one Cray Y-MP processor is in agreement with findings by other authors [10]. Going to lattice sizes of $3 \cdot 2^{16}$ and $5 \cdot 2^{16}$ the speed was down to 36 and 10 SU/microsecond, respectively. This decrease in speed is mainly due to the CSHIFT function. When the Fortran compiler is released, as version 1.0 this might look different. We used version 0.6 for the test.

**Listing 2:** *An implementation of the da Silva-Herrmann algorithm for a vector computer. Any extensions to Fortran 77 are in the syntax of the Cray Fortran CFT77.*

```
1  c       systematic investigation of 1-dim cellular automata
2          PARAMETER (mdim=32000,ibit=64,mdimp1=mdim+1,ldim=ibit*mdim)
3          DIMENSION nr(0:3),n(0:mdimp1),nf(mdim),nh(mdim,0:3),ns(mdim)
4          DATA mt,iseed,pstart/200,13579,0.5/
5          CALL RANSET(2*iseed-1)
6          DO 10 j=1,mdim
7    10    ns(j)=0
8          DO 11 ii=1,ibit
9            DO 11 j=1,mdim
10   11      ns(j)=SHIFT(ns(j).OR.IFIX(pstart+1.0-RANF()),1)
11         DO 2 ica=0,15
12           DO 3 k=0,3
13   3       nr(k)=1.AND.SHIFT(ica,ibit-k)
14         DO 4 j=1,mdim
15   4     n(j)=ns(j)
16  c====== end of initialization, start of time development =======
17         DO 5 it=1,mt
18           n(0)=SHIFT(n(mdim),ibit-1)
19           n(mdimp1)=SHIFT(n(1),1)
20           DO 60 j=1,mdim
21             nh(j,0)=.NOT.n(j+1).AND..NOT.n(j-1)
22             nh(j,1)=      n(j+1).AND..NOT.n(j-1)
23             nh(j,2)=.NOT.n(j+1).AND.     n(j-1)
24             nh(j,3)=      n(j+1).AND.    n(j-1)
25   60        nf(j)=0
26           DO 61 ki=0,3
27             IF (nr(ki).EQ.0) GOTO 61
28             DO 62 j=1,mdim
29   62          nf(j)=nf(j).OR.nh(j,ki)
30   61      CONTINUE
31  c======== end of simulation, start updating and analysis =======
32           m=0
33           DO 7 j=1,mdim
34             n(j)=nf(j)
35   7         m=m.OR.nf(j)
36   5       IF (m.EQ.0) GOTO 2
37   2     WRITE (6,*) ica,m,mt,ldim,iseed,pstart
38         END
```

**Listing 3:** *An implementation of the cellular automata program for a parallel computer. The Fortran syntax is for the Connection Machine 2 of Thinking Machines Corp..*

```fortran
 1  C        systematic investigation of 1-dim cellular automata
 2           PARAMETER (ldim=2**21)
 3           LOGICAL, ARRAY(1:ldim) :: n,ns,nle,nri,nf
 4           REAL,    ARRAY(1:ldim) :: ranf
 5           INTEGER, ARRAY(0:3)    :: nr
 6           DATA pstart,mt,iseed/0.5,1000,1/
 7           CALL CMF_RANDOMIZE(2*iseed-1)
 8           CALL CMF_RANDOM(ranf,0.0)
 9           ns=ranf.LE.pstart
10           DO 2 ica=0,15
11              iica=ica
12              DO 3 k=0,3
13                 nr(k)=MOD(iica,2)
14     3         iica=iica/2
15              n=ns
16  C====== end of initialization, start of time development =======
17              DO 5 it=1,mt
18                 nri=CSHIFT(n,DIM=1,SHIFT=1)
19                 nle=CSHIFT(n,DIM=1,SHIFT=-1)
20                 nf=.FALSE.
21                 IF(nr(0).EQ.1) nf=nf.OR.(.NOT.nri.AND..NOT.nle)
22                 IF(nr(1).EQ.1) nf=nf.OR.(     nri.AND..NOT.nle)
23                 IF(nr(2).EQ.1) nf=nf.OR.(.NOT.nri.AND.     nle)
24                 IF(nr(3).EQ.1) nf=nf.OR.(     nri.AND.     nle)
25  C======== end of simulation, start updating and analysis =======
26                 n=nf
27                 m=COUNT(n)
28     5         IF (m.EQ.0) GOTO 2
29     2      WRITE (6,*) ica,m,mt,ldim,iseed,pstart
30           END
```

## 3. Classification Results

The results for classification of the cellular automata on the five different lattices are shown in Table 1. A dash in the table means that no member in that class exists. A zero means that members must exist, but the number is too small to show up. For example the results with coordination number 4 contain the results for coordination number 2, because the plane can be decomposed into decoupled rows and columns. All results depend more or less monotonically on the coordination number. The most surprising finding is that the results for the two lattices with coordination number 6 are the same within the errorbars. The results do not depend on the geometry of the lattice.

The damage spreading results are shown in Table 2. The introduction of the subclasses $b_0$ and $b_c$ to indicate whether the damage is stationary in space or moves around changed the analysis quite a bit. The two cases can be separated by looking at the size of the damage upon arrival at the surface of the system. The typical size of the damage, when a real class $c$ member reaches the surface is approximately between 0.25 and 0.45. But in the cases, which we count as class $b_c$, the size of the damage is below 0.01. Another possible mechanism of

**Table 1**: *The classification results for the different lattices. The results for coordination number 4 are from Ref. 1 and 2. The abbreviations hc and tri stand for honeycomb and triangular lattice.*

|       | 2 (1-dim.) | 3 (hc.) | 4 (2-dim.) | 6 (tri.) | 6 (3-dim.) |
|-------|------------|---------|------------|----------|------------|
| $a$   | 0.250      | 0.156   | 0.05       | 0.0065   | 0.0064     |
| $b_0$ | 0.125      | 0.063   | -          | 0.0087   | 0.0084     |
| $b_c$ | 0.500      | 0.422   | -          | 0.0003   | 0.0003     |
| $c$   | 0.125      | 0.359   | -          | 0.9845   | 0.9850     |

**Table 2**: *The damage spreading results for the different lattices. The results for coordination number 4 are from Ref. 1. The abbreviations hc and tri stand for honeycomb and triangular lattice.*

|   | 2 (1-dim.) | 3 (hc.) | 4 (2-dim.) | 6 (tri.) | 6 (3-dim.) |
|---|------------|---------|------------|----------|------------|
| 0 | 0.125      | 0.055   | 0.028      | 0.0034   | 0.0039     |
| 1 | 0.125      | 0.055   | 0.028      | 0.0034   | 0.0039     |
| 2 | -          | -       | 0.001      | 0.0002   | 0.0000     |
| 3 | -          | 0.047   | 0.007      | 0.0001   | 0.0000     |
| 4 | 0.125      | 0.047   | 0.021      | 0.0017   | 0.0014     |
| 5 | 0.375      | 0.223   | 0.150      | 0.0001   | 0.0000     |
| 6 | 0.250      | 0.574   | 0.765      | 0.9911   | 0.9908     |

detection is the requirement that the damage is at the surface and at the initially damaged line at the same time.

When we studied the one-dimensional damage spreading, we flipped one spin in the middle of the chain. For six of the 16 rules the outcome depends on the initial condition. Often the damage disappears after one time step. To overcome this we flipped 64 spins equally separated from each other (the distance was a few hundred lattice distances). Now the results did not depend on the initial configuration anymore. To overcome this difficulty we flipped in higher dimensions a whole line of spins. But doing this we introduced a preferred direction into the lattice. This also influenced the results. Assume the rule that does a XOR of the left and right neighbor and ignores the upper and lower neighbor. The plane is decomposed into independent horizontal rows. The outcome of this rule depends on the orientation of the initial damage. If it is horizontal the damage heals out after one time step, if the orientation is vertical the damage spreads horizontally. Nevertheless this rule belongs to class $c$.

The classification results should be compared to new mean field classification results for the square lattice [11]. The author find seven classes of different asymptotic behavior characterized by fixed points and limit cycles of period two, and not the longer periods of class 6, which characterize the majority of cases for coordination number 6.

## 4. Acknowledgements

I thank D. Stauffer for many fruitful discussions and M.A. Novotny for help with the Connection Machine. This research is partially supported by the Florida State University Computations Research Institute which is partially funded by the U.S. Department of Energy through Contract No. DE-FCO5-85ER250000.

## 5. References

1) U. Frisch, B. Hasslacher, and Y. Pomeau, Phys. Rev. Lett. **56**, 1505 (1986).
2) M. Kaufman, J. Urbain, and R. Thomas, J. Theor. Biol. **114**, 527 (1985).
3) S. Wolfram, Rev. Mod. Phys. **55**, 644 (1983).
4) N.H. Packard and S. Wolfram, J. Stat. Phys. **38**, 901 (1985).
5) D. Stauffer, Physica A **157**, 645 (1989).
6) S.S. Manna and D. Stauffer, Physica A, **162**, 176 (1990).
7) R.W. Gerling, Physica A **162**, 187 (1990).
8) R.W. Gerling, Physica A **162**, 196 (1990).
9) L.R. da Silva and H.J. Herrmann, J. Stat. Phys **52**, 463 (1988).
10) S.C. Glotzer, D. Stauffer, and S. Sastry, Physica A, **164**, 1 (1990).
11) Z. Burda, J. Jurkiewicz, and H. Flyvbjerg, preprint for J. Phys. A (1990).

# Simulation Studies of Classical and Non-classical Nucleation

*W. Klein*

Physics Department and Center for Polymer Physics, Boston University,
Boston, MA 02215, USA

We present the results of Monte Carlo studies of nucleation in Ising models with several interaction ranges. We find that for systems with short range interactions the classical theory of nucleation correctly describes the simulation results. However for systems with long range interactions undergoing deep quenches, the nucleation process is not described by the classical theory. We also present the results of simulations done to test the sensitivity of the critical nucleation droplet to various forms of perturbation.

## 1. Introduction and Phenomenology

Nucleation is one of the most common processes in nature, occurring in phenomena as diverse as protein folding, fracture and alloy formation. Controlling this process, e.g. being able to initiate or stop it, is important in several areas of material science and technology. However, in order to be able to control the process, information about nucleation must be obtained on a microscopic scale.

Unfortunately, experimental techniques available to probe nucleation cannot provide information about objects evolving on a milli-second time scale with a linear dimension of a few hundred angstroms. This leaves us with simulation, Monte Carlo or molecular dynamics, as virtually the only source of microscopic information about nucleation and the object that initiates the process, the critical droplet.

The precise nature of the information one needs to obtain control over the nucleation process is not obvious. However one of the first questions one would naturally ask is; What is the structure of the critical droplet? Until relatively recently it was thought that the droplet structure was quite simple. Several arguments generally lumped under the rubric of classical nucleation theory [1] suggest a droplet with the following characteristics. 1. The critical droplet appears as a fluctuation about metastable "equilibrium". 2. There is no interaction between droplets, that is the droplets are isolated. 3. The interior of the droplet is similar to the stable phase in that the density and free energy are close to the stable phase values. 4. There is a well defined surface between the droplet interior and the surrounding metastable phase with a well defined surface tension. Moreover the surface tension is roughly constant with the quench depth (i.e. distance from the coexistence curve.)

With these assumptions the free energy cost of a droplet of size $r$ is

$$\Delta F = -|\Delta f| r^d + \sigma r^{d-1} , \qquad (1)$$

where $\Delta f$ is the free energy density difference between the interior of the droplet and the metastable phase, $\sigma$ is the surface tension and $d$ the dimension of space. The radius of the critical droplet, $r_c$ can be obtained from differentiating eq. 1 and setting the derivative

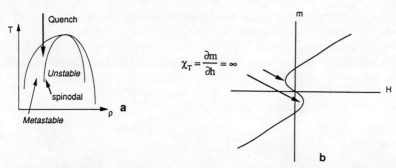

**Fig. 1a** Schematic plot of the coexistence curve and the spinodal for a system with long range interactions. The spinodal separates the metastable from the unstable region.

**Fig. 1b** Schematic plot of the magnetization vs. the magnetic field for a long range interaction system and temperature below $T_c$. The spinodal points are points of divergent susceptibility.

equal to zero. This implies $r_c = (\frac{d-1}{d})\frac{\sigma}{\Delta f}$. Droplets smaller than $r_c$ lower the free energy by shrinking and those larger than $r_c$ by growing.

The probability $p$ of the occurrence of a critical droplet is

$$p = \exp\{-\beta\frac{\Delta F_c}{K_B T}\} = \exp\{-\gamma\frac{\sigma^d}{(\Delta f)^{d-1}}\} \tag{2}$$

where $\gamma$ is a constant that depends on $d$, $K_B$ is Boltzman's constant and $T$ is the temperature. Since the droplets are assumed to be non interacting the probability of a critical droplet is proportional to the nucleation rate, the number of critical droplets per unit volume per unit time. In addition the assumption of isolated droplets allows us to associate the inverse of the probability $p$ with the lifetime of the metastable state. Therefore as the system is quenched further from the coexistence curve and $\Delta f$ increases, the lifetime of the metastable state decreases. When the lifetime is short enough, i.e. when $\gamma\sigma^d/(\Delta f)^{d-1} \sim 2-4$ the metastable state ceases to exist. Specifically the lifetime has become of the order of the relaxation times in the metastable state and hence metastability as quasi-equilibrium is no longer possible. At such quench depths one reaches the so called limit of metastability which is a smeared out crossover region rather that a sharp line.

The prediction for the nucleation rate in eq. 2 has been measured in Ising models with nearest neighbor interactions [2] and found to be correct. Moreover, the structure of the critical droplet in these simulations is consistent with the assumptions of the classical theory. [3] There are however, two problems. First, the classical theory can explain some experimental (as opposed to simulation) measurements of nucleation rates but not all [4]. The second problem requires some discussion.

Suppose we simulate an Ising model with a long range rather than a nearest neighbor interaction. In the limit of infinite range it is known [5] that the description of the system, including the metastable state is given by mean-field theory. For finite but long range interactions the mean-field theory becomes a better approximation the longer the interaction range. [6] In particular the limit of metastability sharpens up as the interaction range increases and in the infinite range limit becomes the spinodal of the mean field theory [6-8] (also see fig. 1).

This result presents serious problems for the classical theory of nucleation. The spinodal is a critical point with a diverging susceptibility. In particular, as with other critical points, the surface tension vanishes as the spinodal is approached. This clearly violates one of the assumptions of the classical theory and would also imply that the critical droplet is not the compact object one sees in nearest neighbor Ising models. If this is true, then anyone doing a simulation test of nucleation near the spinodal has the additional difficulty of not knowing what objects to count as critical droplets. In order to gain some understanding of droplet structure near the spinodal we need a more sophisticated theoretical approach than the one presented above.

## 2. Theoretical Background

To understand the nucleation process in the neighborhood of the spinodal we will need more powerful theoretical techniques. In order to present these techniques we begin by revisiting the classical theory. Various researchers approached the nucleation problem with what is often called field-theory techniques [9] and were able to reformulate the classical nucleation theory in a more rigorous form. The theory reached its present stage of evolution with the work of Cahn and Hilliard [10] and Langer [11,12] and it is this approach we adopt and generalize to treat nucleation near the spinodal. In this theory the critical droplet is related to a saddle point in the free energy surface. We begin with an effective Hamiltonian of the Landau-Ginsburg [9] form.

$$-\beta H(\psi) = \int d\vec{x} \{(\frac{R}{2}\nabla\psi(\vec{x}))^2 - |\epsilon|\psi^2(\vec{x}) + \gamma\psi^4(\vec{x}) - h\psi(\vec{x})\} \quad , \tag{3}$$

where $\epsilon = 1 - T/T_c$, $\gamma$ is a constant independent of $\epsilon$ and $h$, $h$ is the magnetic field and $\psi(\vec{x})$ is the local order parameter. The range of interaction $R = \int d\vec{r} r^2 v(r)$ where $v(r)$ is the potential of interaction. We will always assume $T < T_c$.

The partition function $Z$ is then

$$Z = \int \delta\psi \exp[-\beta H(\psi)] \tag{4}$$

In order to compute the integral in eq. 4 we will assume that we can obtain a good approximation by using the steepest descent method. This means that we restrict our consideration to low temperature or large $R$. We then search for saddle points, that is solutions to the "Euler-Lagrange" equation obtained from the functional derivative of the Landau-Ginsburg Hamiltonian in eq. 3.

$$-R^2\nabla\psi(\vec{x}) - 2|\epsilon|\psi(\vec{x}) + 4\gamma\psi^3(\vec{x}) - h = 0 \quad . \tag{5}$$

We associate the spatially uniform solutions of eq. 5, of which there are 3, two stable and one unstable, with the metastable and stable state. The metastable one of course has the higher free energy. We can now examine fluctuations about the metastable state by looking at the region in function space close to the metastable saddle point independent of the sign of $h$. This amounts to an analytic continuation of the free energy to the metastable region. [11] The fluctuation about the metastable state responsible for nucleation is associated with a spatially non-uniform solution of eq. 5, that is a saddle point. [10-12] There is a detailed discussion of this point in reference [12] however it should seem reasonable from the discussion in section 1. Therefore the solution of eq. 5 is the order parameter profile of the critical droplet.

Near the coexistence curve, i.e. $h \sim 0$

$$\psi(\vec{x}) = \psi_o \tanh\{a(x - x_o)\} \ , \tag{6}$$

where $x = |\vec{x}|$, $\psi_o$ is the magnitude of the spatially uniform order parameter in the metastable state, $a$ is a constant that depends on $R, \epsilon$ and $\gamma$ and $x_o$ is the (arbitrary) location of the surface of the droplet. We have imposed the boundary conditions that $\psi(x = -\infty) = \psi_o$ and $d\psi(x)/dx = 0$ at $x = \infty$.

The droplet profile in eq. 6 allows one to calculate the nucleation rate. Here however we are interested in droplet structure and we refer the interested reader to references [11] and [12] where the method for calculating nucleation rates can be found.

The profile in eq. 6 is consistent with the assumptions of classical nucleation theory. The density in the interior is the same as the stable phase and there is a sharp interface between the droplet interior and the surrounding metastable phase. This gives us some confidence that the saddle point method can produce a description of the critical droplet. We now want to adapt this method to nucleation near the spinodal.

To locate the spinodal we return to eqs. 3 and 4. Since the spinodal is a well defined concept only in the mean-field ($R \to \infty$) limit we can equate the effective Hamiltonian in eq. 3 with the free energy. As discussed above the spinodal is associated with a divergent susceptibility which can be cast in the form $\partial^2 F(\psi)/\partial \psi^2 = 0$ where

$$F(\psi) = -|\epsilon|\psi^2 + \gamma\psi^4 - h\psi \tag{7}$$

For a fixed temperature the spinodal can now be located from the equations

$$\frac{\partial^2 F(\psi)}{\partial \psi^2} = \frac{\partial F(\psi)}{\partial \psi} = 0 \ , \tag{8}$$

where the first derivative equal to zero guarantees that the spinodal point is associated with a free energy minimum.

Writing the spinodal value of $\psi$ obtained from eq. 8 as $\psi_{sp}$ we define a new order parameter

$$\phi(\vec{x}) = \psi(\vec{x}) - \psi_{sp} \ . \tag{9}$$

Inserting eq. 9 into eq. 3 we obtain an expression for the effective Hamiltonian as a functional of $\phi(\vec{x})$

$$-\beta H(\phi) = R^d \int d\tilde{x}^d \{\frac{1}{2}[\nabla\phi(\tilde{x})]^2 + \Delta h \phi(\tilde{x}) - \alpha\phi^3(\tilde{x}) + \gamma\phi^4(\tilde{x})\} \ , \tag{10}$$

where $\tilde{x} = x/R$ and it is understood that the gradient is with respect to the scaled coordinates and we have assumed that $\phi(\vec{x}) = \phi(x)$ The parameter $\alpha = (6|\epsilon|\gamma)^{1/2}$ and $\Delta h = h_s - h$ where $h_s$ is the value of the magnetic field at the spinodal.

We now restrict our consideration to the region near the spinodal, that is, $\Delta h << 1$. In this region we would expect, and we will see below, that $\phi(\tilde{x}) << 1$. Since $\alpha$ remains finite as $\Delta h \to 0$ we can neglect the $\phi^4$ term in the Hamiltonian (eq. 10).

For the critical droplet we follow the same procedure that was outlined in the classical case. In particular the droplet profile is the solution of

$$-\nabla^2 \phi(\tilde{x}) + \Delta h - 3\alpha\phi^2(\tilde{x}) = 0 \ . \tag{11}$$

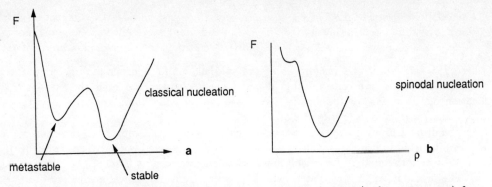

**Fig. 2** Two schematic plots of the free energy vs. the density (order parameter) for a shallow quench (a) and a deep quench (b).

It is simple to solve this equation in one dimension [7] and it has been solved numerically in three dimensions. [13] For our purposes however, we can obtain the information we need from scaling arguments.

It can be seen by substitution that the solution of eq. 11 is of the form

$$\phi(\tilde{x}) \sim (\Delta h)^{1/2} \bar{\phi}[x/R(\Delta h)^{-1/4}] \qquad (12)$$

which justifies neglecting the $\phi^4$ term in eq. 10. This form of the solution indicates that the critical droplet has a linear dimension of the correlation length $\xi$ which diverges at the spinodal critical point [14] as $\xi \sim R(\Delta h)^{-1/4}$. From eq. 12 we also see that the droplet has an amplitude (difference in the order parameter between the interior of the droplet and the metastable state) that vanishes with a critical exponent associated with the thermodynamic order parameter at the spinodal. [14]

Finally if we insert the solution in eq. 12 into eq. 10 we obtain the nucleation barrier to the critical droplet [7]

$$-\beta H(\phi_{crit}) \sim R^d [\Delta h]^{\frac{3}{2} - \frac{d}{4}} \qquad (13)$$

which is of the form $\xi^d [\Delta h]^{2-\alpha}$, i.e. the form of the free energy barrier to fluctuations associated with singularities at the thermal critical point.

Summarizing: The critical droplet has the linear dimension of the correlation length. The order parameter difference between the droplet interior and the surrounding metastable state goes to zero as the spinodal is approached in a manner identical to the vanishing of the thermodynamic order parameter at the spinodal. Finally, the free energy cost of the critical droplet is the same form as the free energy cost associated with a fluctuation the size of the correlation length near the spinodal. Taken together these three results strongly suggest that for deep quenches in systems with long range interactions the critical droplet that initiates the metastable state decay is a fluctuation associated with the thermal critical point called a spinodal. Moreover, nucleation near the spinodal is not a mechanism by which there is "tunneling" from the bottom of the metastable well to the bottom of the stable well as in the classical theory. In "spinodal nucleation" the system "tunnels" from the metastable well to the top of the stable well using a critical fluctuation. (cf fig. 2)

This result partially answers the question posed at the end of section 1. Clearly near the spinodal the nature of the critical droplet will not be the same as in the classical theory of nucleation. Critical phenomena fluctuations are not associated with compact objects such as the those seen in nucleation near the coexistence cvurve. In fact they are known at Ising critical points to be fractals. [15] This pesents us with a serious problem if we wish to verify the theoretical predictions with simulation and an interesting question of importance to the problem of controlling the nucleation process.

The question of the droplet structure near the spinodal will be addressed in the next section. However here I would like to make two points about the nucleation barrier (eq. 13). First, for dimensions less than 6 the nucleation barrier will decrease as $\Delta h \to 0$. In order to get close to the spinodal, and remain in the metastable state, we must have the nucleation barrier relatively high. This implies that the range of interaction $R$ must be large. This is why spinodals are not a useful concept in short range interaction systems and why the homogeneous nucleation process is strongly material dependent.

The second point I wish to mention is that for $d > 6$, eq. 13 implies that there is a spinodal for short range interaction systems. This prediction has been tested by Monte Carlo simulations on nearest neighbor Ising models [16] and appears to be correct.

## 3. Percolation Mapping

In order to test the theoretical predictions in section 2 we need to have a precise definition or prediction for the structure of the droplet. It is not sufficient to say that the critical droplet is fluctuation associated with a critical point or that it is a fractal. In order to solve this problem we use the result that the critical droplet is a critical point fluctuation and use the idea that critical points can be mapped onto percolation problems [15].

We begin by noting that Kasteleyn and Fortuin showed that random bond percolation problems could be described using the $q \to 1$ limit of the $q$ state Potts model [17]. Specifically we consider a lattice with a spin at each site that can be in one of $q$ states. The interaction between pairs of spins is of the form

$$-\beta H = J \sum_{ij}(\delta_{\sigma_i \sigma_j} - 1) - \sum_i h(\delta_{\sigma_i,1} - 1) \tag{14}$$

where $h$ is an external magnetic field, $\delta_{\sigma_i \sigma_j}$ is the Kronecker delta and the $\sigma_i$ specifies the state of the $i^{th}$ Potts spin. Note that the sum is **not** restricted to nearest neighbor pairs. If the free energy formed from the Hamiltonian in eq. 14 is differentiated with respect to $q$ and the limit $q \to 1$ is taken, the generating function for the random bond percolation problem is obtained.

This result can be easily generalized to treat correlated site - random bond percolation by considering the Hamiltonian

$$-\beta H = J \sum_{ij}(\delta_{\sigma_i \sigma_j} - 1)n_i n_j - \sum_i (\delta_{\sigma_i} - 1)n_i + H_I(n_i) \ , \tag{15}$$

where $n_i$ is an occupation number which is either one (site occupied) or zero (site empty). The term $H_I$ specifies the interaction between the $n_i$. This is the so called dilute $q$ state Potts model.

If the same procedure of differentiation with respect to $q$ and the limit $q \to 1$ is followed, then one can show that the Hamiltonian in eq. 15 describes a site-bond percolation problem where the sites are distributed according to the Boltzman factor with the Hamiltonian $H_I$, and the bonds are distributed at random between occupied sites with a probability $p_B = 1 - \exp(-J)$.

Finally, if $H_I$ is chosen to be the lattice gas Hamiltonian

$$-\beta H_I = K\sum_{ij} n_i n_j - \Delta \sum_i n_i \ , \qquad (16)$$

where $K$ is the interaction constant and $\Delta$ the chemical potential and $J$ in eq. 15 is fixed at $K/2$, then the critical point of the lattice gas can be shown to be a percolation transition of the correlated site - random bond model where the bond probability $p_B = 1 - \exp(-K/2)$. Moreover all the critical exponents that characterize the percolation transition are identical to those of the lattice gas critical point. For example, the correlation exponent $\nu$ of the lattice gas model is equal to the connectedness length exponent of the percolation model [15].

Before considering the spinodal there are two points of a technical nature that should be mentioned. First, if instead of the lattice gas Hamiltonian one chose the Potts interaction with $q = 2$ for $H_I$ then one would set $J = K$ to make the thermal critical point and the percolation transition coincide. Second, if one defines bonds between holes (i. e. $1 - n_i = 1$) as well as particles ($n_i = 1$) then the mapping at the critical point is complete in that critical amplitudes as well as critical exponents are the same in the thermal and percolation transitions.

In order to treat the spinodal transition we use the mean field formulation of the dilute $q$ state Potts model where $H_I$ has been chosen to be the $q = 2$ Potts Hamiltonian. [18] In the Landau - Ginsburg form the free energy per spin is given by

$$\frac{F_P}{N} = -\frac{r_1}{2}(q-1)\psi_P^2 - (q-1)h_P\psi_P + \frac{1}{3}w_1 q(q-1)(q-2)\psi_P^3 + \frac{1}{2}w_2 q(q-1)\psi\psi_P^2 + F(\psi), (17)$$

where $F(\psi)$ is the Ising or Potts free energy per spin given in eq. 7 and obtained from $H_I$ in the limit of long range interactions (i.e. $R \to \infty$), $\psi_P$ is the Potts order parameter obtained from the Potts interaction that will give rise to $p_B$ and $h_P$ is the Potts magnetic field. The terms $r_1, w_1$ and $w_2$ are parameters obtained from the transformation from the discreet dilute $q$ state Potts model of eq. 15 to this continuum version. [18] Note that the term proportional to $w_2$ is the only coupling between $F(\psi)$ and the Potts percolation term.

In order to obtain the generating function for the percolation problem $f_P$ we differentiate eq. 17 with respect to $q$ and set $q = 1$.

$$f_P = \frac{1}{2}(-r_1 + w_2\psi)\psi_P^2 - h_P\psi_P + \frac{1}{3}w_1\psi_P^3 \ . \qquad (18)$$

The percolation transition is found by the usual method of setting $h_P = 0$ and requiring that $r_1 = w_2\psi$. [18] In order to put this condition in the form usually found in the literature for the spinodal transition [19,20] we first put $\psi = \psi_{sp}$ the value of the order parameter at the spinodal.( see eqs. 8 and 9). Next we insert the values for $w_2 = \frac{1}{2}c^3 J K^{1/2}$ and $r_1 = c(1 - cJ/2)$ where $c$ is the lattice coordination number and the lattice parameter has been set equal to one.[18] The final step we need to take is to note that as the temperature goes to zero ($K \to \infty$) the value of the magnetization per spin $m$ at the spinodal goes to 1. The value of $\psi_{sp}$ however is given by

$$\psi_{sp} = \pm \frac{\sqrt{(cK-1)c}}{c^2 K} \ . \qquad (19)$$

As $K \to \infty$ $\psi_{sp}$ goes to zero. This implies that

$$m = cK^{1/2}\psi_{sp} \ . \qquad (20)$$

The density of occupied sites $\rho$ is related to $m$ by $\rho = (1+m)/2$. Combining all these results we obtain for our percolation condition $J = K(1-\rho)$. From the discussion following eq. 15 this implies a bond probability for the percolation mapping onto the spinodal line of

$$p_B = 1 - \exp[2K(1-\rho)] \ . \tag{21}$$

Again for those interested in the details this mapping results from using the Potts form for $H_I$. If the lattice gas form was used there is no factor of 2 in the exponential.

We now have a precise description of the critical droplet close to the spinodal and are in a position to test the theory and ask some interesting questions about the behavior of the critical droplets.

## 4. Monte Carlo Simulations of Nucleation in Ising Models

### 4.1 Cluster Structure

The first set of simulations I will discuss is the test of the field theory predictions of section 2. We simulate an Ising model with long range interactions. Let us be specific and define our interaction as follows: Around each spin on the lattice we construct a square of side $l$ in $d = 2$ or a cube of side $l$ in $d = 3$ with the spin at the center. The center spin interacts with all spins within the square or cube with an interaction strength $K = K_o/K_B T l^d$ and has zero interaction with spins outside. After some number of Monte Carlo steps (which is to a large extent arbitrary) we stop the simulation and throw bonds at random between pairs of occupied (spin up say) interacting sites with a probability given by eq. 21. From the mapping presented in section 3 these percolation clusters are the geometric realization of the critical phenomena fluctuations and, by the saddle point arguments presented in section 2, the largest of these clusters is the critical nucleation droplet.

In order to locate the droplet we [20] plot (in $d = 2$) the x and y coordinates of the center of mass of the largest cluster. A plot of the coordinates of the center of mass as a function of time is presented in fig. 3 for an interaction range such that one spin interacts with $l^2 = q = 684$ neighbors and a quench depth such that $(\Delta h)^{1/2} \sim 0.1$. The linear

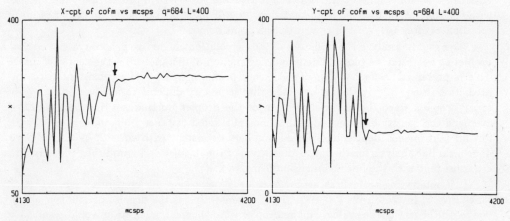

**Fig. 3** The x and y coordinates of the center of mass of the largest cluster as a function of time in Monte Carlo time steps.

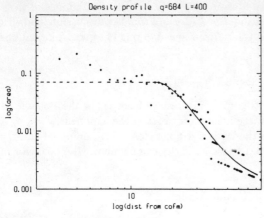

**Fig. 4** Density profile of the critical droplet for the number of neighbors $q = 684$ and quench depth $(\Delta h)^{1/2} = 0.1$

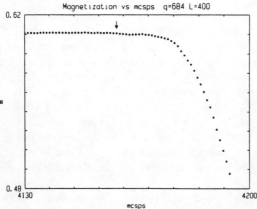

**Fig. 5** Magnetization per spin vs. time. The arrow indicates the time of nucleation for the same quench as in Fig. 4.

system size L = 400 spins. Initially the center of mass location fluctuates wildly due to the fact that clusters grow and decay so that the identity of the largest cluster changes with time. At some time $t_N$ the center of mass stablizes. We will provisionally identify $t_N$ with the time of nucleation. Note that there is still some fluctuation in the center of mass coordinates after $t_N$. We will return to this point below.

We can now analyze the structure of the percolation cluster we propose as the critical droplet at $t_N$. First we plot the density as a function of distance from the center of mass. This is presented in fig. 4 [20]. The density profile appears to be consistent with the prediction in eq. 12. It should be noted that it is very difficult to average profiles over many droplets since the linear dimension of the droplet fluctuates. However, we have averaged the density over several realizations at a fixed distance from the center of mass for different quench depths and obtain numbers consistent with eq. 12. [20]. We have done a similar analysis for the linear dimension of the droplet and found that the numbers are consistent with the linear dimension being $\xi$ [20].

It is interesting to look at a plot of the magnetization per spin $m$ as a function of time (fig. 5). In these runs $\xi \sim 40$ lattice constants, that is the droplet diameter is about one tenth the size of the system. If the droplet were classical, nucleation would result in a noticeable change in $m$. As can be seen in fig. 5 $t_N$, as indicated by the arrow, is about 25 Monte Carlo steps earlier than any noticeable change in $m$.

We have conducted these tests with both the Metropolis algorithm [21] and the Creutz global demon algorithm. [20] The latter conserves the energy. We have noticed no significant difference as far as the droplet structure is concerned. This is consistent with the theory presented in section 2.

We conclude this section by discussing whether $t_N$ is consistent with the nucleation time predicted by the field theory. The theory associates the appearence of a critical droplet with the system being at the top of a saddle point [11,12,14]. If this is true of the system at $t_N$ then if we perturb the system randomly at $t_N$ half of droplets should continue to grow and half decay away. We performed the following simulation. [22] We ran the system described above until the nucleation and growth processes were completed. We then reran it with the same initial conditions and the same random number seed until $t_N$. At $t_N$ we stopped the run, changed the random number seed, and restarted the simulation. Within statistical error we found that in half the runs the original droplet continued to grow and in half the runs the original critical droplet decayed. This would indicate that the stabilizing of the center of mass of the cluster is coincidental with the system reaching the saddle point.

We then proceeded to perform the same tests on nearest neighbor Ising models where the nucleation process is known to be classical. [2] We found that the stablization of the center of mass of the largest cluster occurred 60 to 80 Monte Carlo steps before the saddle point. [22] In fact in the runs we did all of the "center of mass" droplets decayed if we intervened by changing the random number seed.

This would indicate that in classical nucleation there is a growing object that can be identified long before the saddle point is reached and that this is not the case for spinodal nucleation. We conjecture that this is connected with the ramified and hence rather tenuous nature of the critical droplets near the spinodal. It also has interesting implications if one is looking for a way to control nucleation.

## 4.2 Simulation of Early Stage Growth

We turn now to another prediction of the field theory approach. If we want to understand how the critical droplet begins to grow just after it appears, we can linearize the equation of motion around the saddle point solution. [9,12] In model A in the Hohenberg - Halperin classification, [9,23] which would correspond to Metropolis Monte Carlo, the equation of motion for the time dependent order parameter $\psi(x,t)$ is

$$\frac{\partial \psi(x,t)}{\partial t} = -\nabla^2 \psi(x,t) - 2|\epsilon|\psi(x,t) + 4\gamma\psi^3(x,t) - h \quad , \tag{22}$$

where the right hand side of eq. 22 is identical to the left hand side of eq. 5. Equation 22 is linearized by inserting $\psi(x,t) = \psi_o(x) + u(x,t)$. The function $\psi_o(x)$ is the critical droplet profile i.e. a solution of eq. 5 or eq. 11 and the function $u(x,t)$ is assumed to be small. Expanding and keeping only linear terms in $u(x,t)$ results in an equation similar in form to the Schrödinger equation which can be solved for various quench depths. This has been done numerically and the profile for the eigenvector with a negative eigenvalue are plotted in fig. 6.

The interpretation of the eigenvector associated with the negative eigenvalue [12] is that at a saddle point there is one direction in function space that specifies the path of steepest descent away from the saddle point. This will be given by the eigenvector of the equation of motion including the conservation laws. In model A (Metropolis) there are no conservation laws. The situation is more complicated in model B (Kawasaki dynamics) for

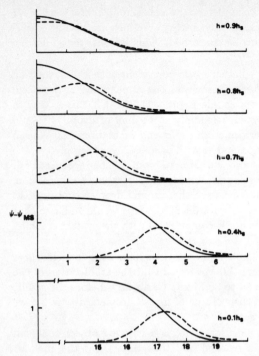

**Fig. 6** Critical droplet and eigenvector profiles obtained from numerical solutions of the field theory equations for various quench depths relative to the spinodal at $h = h_s$

example. [24] The direction away from the saddle point is associated with the early stages of growth of the droplet.

Note in fig. 6 that near the coexistence curve the eigenvector peaks on the droplet surface but that in the deep quench near the spinodal the eigenvector peaks in the center of the droplet. The interpretation is that in contrast to classical nucleation where the droplet grows by adding matter to the surface droplets grow initially in spinodal nucleation by filling in.

This interpretation was tested with Monte Carlo and the results are plotted in fig. 7. The log of the size of the droplet (number of spins) is shown vs. the log of the radius of gyration at several times. If the droplet were growing classically one would expect a line with slope $d$, the spatial dimension. Instead one finds that initially the radius of gyration hardly grows at all while the size increases rapidly. This is consistent with the interpretation of the eigenvector as indicating a filling in or compactification in the early growth stage.

This result is another indication that the critical droplet near the spinodal is indeed a ramified object.

### 4.3 Perturbations

In this final section I will discuss the results of perturbing the critical droplets to test their sensitivity as a function of the distance from the center of the droplet. The question we tried to answer [22] is; How well does the eigenvector that describes growth near the saddle point describe the decay of droplets that either don't quite make it over the top of the saddle point or that are "pushed" back down the hill toward the metastable minimum?

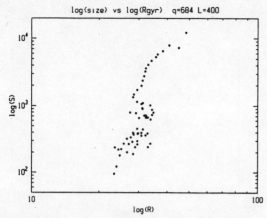

**Fig. 7** Log of the number of spins in the droplet (size) vs. the log of the radius of gyration for various times after nucleation. Each point represents an average over 16 runs.

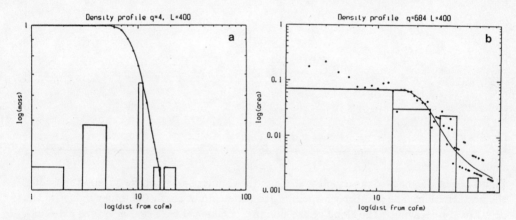

**Fig. 8** Result of perturbation of the critical droplet at various locations in the classical (a) case and the spinodal (b). The heights of the histograms represent the relative frequency of droplet decay after the perturbation.

To answer this question we again used the method of allowing a simulation of a nucleation event to run to completion and then rerunning the system with the same initial condition and random number seed to $t_N$. At $t_N$ we then take the critical droplet and perturb it by removing spins in shells. For example we might, in the spinodal nucleation case, remove all the spins in a disc centered at the center of the droplet with a radius of 10 lattice constants and then in separate runs remove spins in annuli with large radii.

The results of these runs are shown in figs. 8 and 9. The histograms in fig. 8 are superimposed on the droplet profiles in the two cases, classical and spinodal, to indicate where the spins were removed. The height of the tallest histogram is normalized to one so that the height gives the percentage of trials in which removal of the spins in the indicated region of the droplet resulted in that droplet dying. Removal of the spins in the region indicated by the tallest histogram resulted in the droplet dying in all cases.

Note that the region of greatest sensitivity is identical to the location of the peak in the eigenvector in fig. 6 in both the classical and spinodal cases. In fig. 9 is plotted the log

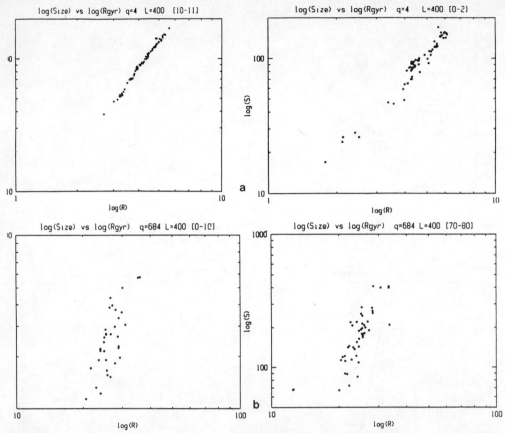

**Fig. 9** Log of the mass of the droplet vs. log of the radius of gyration for several times after the perturbation. The points corresponding to earlier times are in the upper right and the droplets are decaying. The slope of the line in the classical case (a) is two as expected. The plots are virtually identical to the ones for growth. The numbers in brackets indicate the regions from which spins were removed.

of the size of the droplet vs. the log of the radius of gyration as in fig. 7 except here time starts after the perturbation and the droplets are **decaying** The points that correspond to the earliest time have the largest size. In the classical case the line has a slope 2 as predicted by the eigenvector approach. The plot in the spinodal case is virtually identical to the growth plot in fig. 7.

A final point is that the eigenvector in the classical case extends outside the critical droplet (cf. fig 6 for $h = .1h_s$). This is also seen in the simulations of the perturbation in the classical case (fig. 8a). This can be interpreted as the region of enhanced sensitivity extending beyond the droplet in the classical case. This is neither predicted or seen in the spinodal region.

It would seem that the eigenvector correctly describes the response to perturbations about the saddle point.

## 5. Conclusion

We would conclude from these simulations that the nucleation near the spinodal is not described by the classical theory and that the droplets in this region are indeed ramified. These results also indicate that the association of the appearence of a monotonically growing cluster with the nucleation event (i.e. with a saddle point) is only correct in the spinodal region. Finally it appears that the eigenvector approach to early growth will also describe the initial decay of droplets that are pushed away from the saddle point back toward the metastable state.

## Acknowledgements

Several people have collaborated in the work presented here. They include Dieter Heermann, Chris Unger, Tane Ray, Liza Monette, Dietrich Stauffer, Antonio Coniglio, Martin Zuckermann and Richard Harris. In addition I have also benefitted from collaboration on related work and several discussions with J. Yang, Harvey Gould, Ray Mountain, Francois Leyvraz Mark Novotny, Per Rikvold and John Rundle. Finally I have benefitted from discussions with Kurt Binder, Martin Grant, Rashmi Desai and John Cahn. I would also like to acknowledge the support of the ONR and the Academic Computing Center at Boston University.

## 6. References

1. *Nucleation* ed. A. C. Zettlemoyer, Dekker, New York (1969)
2. D. Stauffer, A. Coniglio and D. W. Heermann, Phys. Rev. Lett. **49**, 1299 (1982)
3. L. Monette and W. Klein (in preparation)
4a. J. L. Schmitt, G. W. Adams and R. A. Zalabsky, J. Chem. Phys. **77**, 2089 (1982)
4b. J. L. Schmitt, R. A. Zalabasky and G. W. Adams, J. Chem. Phys. **79**, 638 (1983)
4c. G. W. Adams, J. L. Schmitt and R. A. Zalabasky, J. Chem. Phys. **81**, 5074 (1984)
5. J. L. Lebowitz and O. Penrose, J. Math. Phys. **7**, 98 (1966)
6. D. W. Heermann, W. Klein and D. Stauffer, Phys. Rev. Lett. **49**, 1261 (1982)
7. W. Klein and C. Unger, Phys. Rev. **B28**, 445 (1983)
8. K. Binder, Phys. Rev. **A29**, 341 (1984)
9. See J. D. Gunton, M. San Miguel and P. S. Sahni, in *Phase Transitions and Critical Phenomena,* ed. C. Domb and J. L. Lebowitz (Academic, New York 1983) Vol. 8 and references therein
10. J. W. Cahn and J. E. Hilliard, J. Chem. Phys. **28**, 258 (1958)
11. J. S. Langer, Annals of Phys. **41**, 108 (1967)
12. J. S. Langer, Annals of Phys. **54**, 258 (1969)
13. C. Unger and W. Klein, Phys. Rev. **B31**, 6127 (1985)
14. C. Unger and W. Klein, Phys. Rev. **B29**, 2698 (1984)
15. A. Coniglio and W. Klein, J. Phys. **A13**, 2775 (1980)
16. T. Ray (preprint)
17. P. W. Kasteleyn and C. M. Fortuin, J. Phys. Soc. Japan Suppl. **26**, 11 (1969)
18. A. Coniglio and T. C. Lubensky, J. Phys. **A13**, 1783 (1980)
19. D. W. Heermann and W. Klein, Phys Rev. **B27**, 1732 (1983)

20. L. Monette, W. Klein, M. Zuckermann, M. Khadir and R. Harris, Phys. Rev. **B38**, 11607 (1988)
21. T. Ray and W. Klein (in preparation)
22. L. Monette, W. Klein and M. Zuckermann (in preparation)
23. P. C. Hohenberg and B. I. Halperin, Rev. Mod. Phys. **49**, 435 (1977)
24. L. Monette and W. Klein (in preparation)

# Molecular Dynamics of Slow Viscous Flows

*J.R. Banavar[1], J. Koplik[2], and J.F. Willemsen[3]*

[1] Department of Physics and Materials Research Laboratory, Pennsylvania State University, University Park, PA 16802, USA
[2] Benjamin Levich Institute and Department of Physics, City College of the City University of New York, New York, NY 10031, USA
[3] RSMAS – AMP, 4600 Rickenbacker Causeway, University of Miami, Miami, FL 33149, USA

We use molecular dynamics techniques to study the microscopic aspects of several slow viscous flows past a solid wall, where both fluid and wall have a molecular structure. Systems of several thousand molecules are found to exhibit reasonable continuum behavior, albeit with significant thermal fluctuations. In Couette and Poiseuille flow of liquids we find the no-slip boundary condition arises naturally as a consequence of molecular roughness, and that the velocity and stress fields agree with the solutions of the Stokes equations. At lower densities slip appears, which can be incorporated into a flow-independent slip-length boundary condition. An immiscible two-fluid system is simulated by a species-dependent intermolecular interaction. We observe a static meniscus whose contact angle agrees with simple estimates and, when motion occurs, velocity-dependent advancing and receding angles. The local velocity field near a moving contact line shows a breakdown of the no-slip condition.

## 1. Introduction

The equations of continuum fluid mechanics are incomplete without specification of boundary conditions. For over a century, the no-slip condition that fluid velocity vanishes at a solid surface has successfully accounted for the experimental facts, but several problems remain. The first is that there is no compelling theoretical argument for no-slip. As we shall review below, kinetic theory provides some insight into the question, but cannot deal with the realistic situation — an interacting many-body problem of a dense liquid adjacent to a nearly-rigid molecular solid. Indeed, the no-slip condition is usually presented as an empirical result of 19th century experiments[1,2], and has not been questioned until recently. For most purposes it is not necessary to do so, but a serious problem arises when a contact line separating two immiscible fluids moves along a solid surface[3-5]. The straightforward hydrodynamic analysis of this situation predicts a divergent energy dissipation rate, and slip boundary conditions, among other explanations, have been proposed to deal with the problem. Lastly, closely related questions involving the connection between the microscopic scale and the continuum arise in the description of fluid phenomena in very small systems where the number of molecules is not that large. Examples of current interest include the study of very thin films and surface forces[6] and flow in microporous solids[7].

In this paper we address some of these issues using molecular dynamics (MD) simulations of several cases of fluid flow near a solid boundary. In MD one explicitly computes the detailed molecular motion from the instantaneous positions and velocities and an interaction potential[8-10]. A key feature of our work is the use of a wall with molecular structure. We are thus able to set up a controlled and unprejudiced numerical experiment on a system with specified reasonable microscopic properties, examine the resulting average or large scale behavior, and correlate the results to the continuum description. Compared to other methods there are both advantages and disadvantages to MD. In contrast to laboratory experiments, there are no difficulties with contaminant effects, optical distortion, or microscopic flow visualization. Compared to analytic methods, no assumptions about correlations in intermolecular distribution functions are required. The negative side of MD is its current practical restriction to small systems (up to about 100 Å) and short times ($10^{-9}$ sec or less), with concomitantly large thermal and statistical fluctuations. In consequence, we have taken some pains to verify that continuum behavior is observed in our simulations.

The utility of MD in the study of some fluid mechanical problems has been recognized earlier by a number of authors. In addition to work on channel flows which we shall describe in detail below, interesting calculations of flow past obstacles[11] and Rayleigh-Benard convection[12] have appeared recently. The particular stress of the present paper is rather on using MD as a bridge between the microscopic and macroscopic levels of description. In this sense, our closest precursor is the work of Alder and collaborators[13] on velocity correlations and Brownian motion in hard sphere systems, where extrapolation of ``micro-hydrodynamics'' to the continuum is examined. Other accounts of our work have appeared in the literature.[14]

## 2. The No-Slip Condition

The no-slip condition states that the velocity of a viscous fluid at a solid surface vanishes. The vanishing of the normal component is not controversial, as this is essentially the definition of a solid surface as one which confines the fluid. The vanishing of the tangential component is not obvious at all, however. While one would naturally expect the solid to exert a frictional force on fluid moving past and tend to slow it down, it is unclear why such friction would bring the fluid exactly to rest. The first quantitative theoretical study of the no-slip condition is that of Maxwell[15], who considered a dilute gas of molecules in the presence of an idealized solid wall. A molecule colliding with the wall was assumed to undergo diffuse reflection with probability f or specular reflection with complimentary probability 1-f. In specular reflection the colliding molecule reverses its velocity normal to the wall, but preserves its tangential velocity, as in an elastic collision with a very massive solid. Diffuse reflection was an idealization of a multiple collision process between a gas molecule and the wall molecules, wherein the former emerges from the wall after adsorption and evaporation with a velocity appropriate to a Maxwellian temperature distribution, as if the wall acted as a heat bath. Maxwell showed that the average tangential velocity u at the wall was related to the shear rate $\partial u/\partial z$, z being the coordinate normal to the wall, by

$$u = \lambda \frac{\partial u}{\partial z} \quad \text{(wall)}, \tag{1}$$

where the "slip length" is given by

$$\lambda = \frac{2}{3}(2/f - 1)l, \tag{2}$$

where l is the mean free path of the gas molecules. Now in a dilute gas, the mean free path can become appreciable and this argument then predicts slip at a solid surface, a phenomenon known to occur for dilute gases in the Knudsen regime[16]. On the other hand, if one (heuristically) applied the argument to a liquid, where the mean free path is comparable to the molecular size, the tangential velocity would be proportional to a microscopic number and, on the macroscopic scale, would effectively be zero.

While for most applications the no-slip condition may be accepted uncritically as a phenomenological rule, recent work on moving contact lines has raised a serious question of principle. The difficulty appears most simply in an analysis

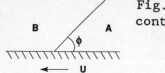

Fig. 1. Definition sketch for a moving contact line.

due to Huh and Scriven[3]. Consider the steady state displacement of fluid A by fluid B along a plane solid surface, shown in Fig. 1 in the rest frame of the meniscus where the solid moves backward with velocity U. In the immediate vicinity of the contact line it is plausible to assume that the meniscus appears flat with some contact angle $\phi$, the flow field is two-dimensional, and the relative velocity of fluid and solid is sufficiently small for the Stokes equations to apply. The critical problem is that a measurable quantity, E, the energy dissipation per unit time per unit length of contact line, has a logarithmic divergence at small r (r is the distance from the contact line). If for purposes of discussion the integration is cut off at a small value $r = \lambda$, then

$$E \sim \log(R/\lambda), \qquad (3)$$

where R is some outer scale of the flow. While the details of this argument may be criticized[3,4], e.g., the normal stress boundary condition is not satisfied, the conclusion that a singularity arises at a moving contact line survives. A fairly general argument for this assertion is given by Dussan and Davis[17], who by experiment and kinematic analysis show essentially that a moving contact line is consistent with the no-slip condition when the fluids undergo a rolling motion. They further show that this motion necessarily gives a multiple-valued velocity which implies a singularity in the force exerted by the fluid on the solid.

A divergence in a measurable quantity usually represents an inconsistency in the assumptions that produced it, and many proposals have been advanced to remedy the moving contact line singularity. In view of the earlier discussion in this paper and the results to be given, we favor relaxing the no-slip condition, but other possibilities exist. First of all, the contact line motion may be unsteady, exhibiting ''stick and slip'' motion[18], and perhaps may be occasionally rapid enough for inertial effects to enter. Secondly, as alluded to above, the interface may not be locally flat and some curvature may be present down to microscopic length scales giving an oscillatory interface[19]. In addition, van der Waals forces should enter at short distances and perhaps remove the singularity[20]. More generally, van der Waals forces may give rise to ''precursor films'' advancing ahead of the bulk meniscus. The no-slip condition at the contact line is then replaced by a matching to the precursor film[5], but it is not clear what happens at the tip of the film.

If one allows slip at a solid surface, the first issue is the precise functional form of the boundary condition. This may not be critical, as Dussan[21] has shown that the macroscopic flow resulting from a variety of particular forms of slip conditions is the same, all giving a finite stress and an energy dissipation rate of the form (3). The question is now the magnitude of the slip length. The simplest possibility is that $\lambda$ is of the order of a molecular size, but another is that the slip length is related to roughness of the solid surface. Various calculations[22] have shown how a slip-free flow over a rough surface can lead to effective slip, perhaps associated with stick-slip motion of the meniscus over the irregularities. One might then say $\lambda$ is of order the roughness size, but other calculations[23] find a more complicated dependence.

To summarize this situation, there is no consensus on the correct resolution of the moving contact line singularity problem. As with all of the issues discussed in this section, macroscopic modeling, kinetic theory, or laboratory experiments do not yield a definite conclusion, and the MD technique seems ideally suited to their elucidation.

## 3. Molecular Dymanics with Walls

We have used a standard molecular dynamics algorithm[8-10] in which a pair of molecules separated by distance $r$ interacts through a 12-6 Lennard-Jones potential

$$V_{IJ}(r) = 4\varepsilon[(r/\sigma)^{-12} - (r/\sigma)^{-6}] + r\delta V, \qquad (4)$$

characterized by energy and distance scales $\varepsilon$ and $\sigma$, respectively. The potential has a repulsive core at $r \sim \sigma$, and the depth of the attractive well is $-\varepsilon$. It is computationally convenient to cut off the potential and treat purely short-range interactions, and we do so at a value $r_c = 2.5\sigma$. The additional correction term $r\delta V$ is inserted so that the force vanishes at the cutoff. The chosen interaction is known to successfully reproduce the properties of liquid argon, if the parameters are chosen as $\sigma = 3.4$ Å and $\varepsilon/k_B = 120°K$, along with a molecular mass m = 40 a.u. The natural time unit in the calculations is then $\tau = \sigma\sqrt{m/\varepsilon} = 2.16 \times 10^{-12}$ sec. In the remainder of this paper, all dimensional quantities given as pure numbers will be understood to be multiplied by an appropriate combination of $\sigma$, $\varepsilon$ and m.

In our computer simulations, the molecular positions are specified to give the instantaneous potential, and Newton's law is numerically integrated using a fifth order predictor-corrector scheme, with a time step of $0.005\tau$. The molecules are initialized on an fcc lattice whose spacing is chosen to obtain the desired density, with initial velocities randomly assigned subject to a fixed temperature. The substance is then allowed to melt. To simulate an infinite system, the system would be placed in a box with periodic boundary conditions on all sides, so that a molecule exiting the box is reinserted on the opposite face. Periodic copies of the molecules are used in the force computation. To study flow in a channel geometry, we have modified the algorithm to provide constraining walls on two sides, while retaining periodicity in the other two directions.

We assign the molecules in the top and bottom two layers of the initial fcc lattice a heavy mass, $m_H = 10^{10} m$, but allow these to move in accord with the equations of motion. In this way collisions between fluid and wall molecules conserve energy, and the walls retain their integrity over the duration of the simulation, although eventually they would disintegrate. In the simplest case, the wall-fluid and fluid-fluid interactions are assumed to be the same.

The phase space trajectory of the collection of the molecules is obtained by monitoring their positions and velocities. The internal energy, the displacement of each molecule from its starting position, the running average temperature of the two walls and of the fluid and the total energy are all monitored as a function of the time. The region between the walls is divided into bins along the z direction (for the immiscible two-fluid simulations below, the channel is binned across the gap as well) to obtain running averages for the density, the mean x, y and z displacements and velocities and their squares, and the nine components of the stress tensor in each of the bins. Typically the number of bins was equal to the twice the number of unit cells of the initial fcc lattice, giving a bin width $O(\sigma)$.

For our initial studies, the density was chosen to be $\rho = 0.8$ and temperature to be 1.2. This should place us in the liquid region of the bulk phase diagram. Most of our runs were carried out with 1536 fluid molecules and 256 molecules for each of the two walls. The system was divided into 12 bins in this case, having an average of 128 molecules per bin. The fluid melts and achieves equilibrium (defined here as a steady state in temperature, pressure, energy, etc.) after several thousand time steps. In Fig. 2 we show an x-z snapshot of instantaneous molecular positions after 5000 time

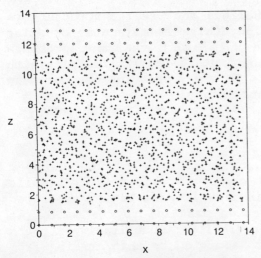

Fig. 2. Instantaneous molecular positions for a liquid in equilibrium between walls at T = 1.2 and $\rho$ = 0.8, where molecular positions for all values of y are superposed.

steps of constant-temperature equilibration, in which molecules at all values of y are superposed. Note first that the wall molecules remain very close to their original lattice sites and that the fluid molecules are indeed confined. Near the walls distinct molecular ordering can be seen: the ordering normal to the walls is exhibited quantitatively in Fig. 3, where the solid line gives the time-averaged density profile as a function of z. This layering effect is consistent both with experiment[24,25] and with earlier MD studies of fluids confined by either artificial or molecular

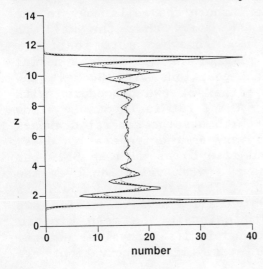

Fig. 3. Average density profile associated with Fig. 2.

boundaries,[26-30] although we believe the precise structure to be model-dependent. In addition, there is molecular ordering in the layer adjacent to the wall, with the most probable location of a fluid molecule being at an fcc lattice site with respect to the wall molecules. This x–y ordering has not been directly observed in experiments as yet.

## 4. Channel Flows of a Single Fluid

We now turn to the motion of a fluid confined between molecular walls, and the first case we consider is Poiseuille flow in the liquid phase. This flow is most easily generated by applying a uniform acceleration parallel to the walls, as if the fluid were falling under gravity. Quantitatively, each molecule experiences a force in addition to that arising from (4) of the form

$$\delta F = mgx .\qquad(5)$$

Because we are dealing with small systems and short times, some care must be taken in order to extract a signal from substantial thermal noise. In Lennard–Jones units $(\varepsilon,\sigma,m)$, typical velocities and forces are $O(1)$, whereas the real, laboratory force of gravity is $O(10^{-13})$, and would have no discernible effect. A much larger value is needed, but we are constrained to simulate low Reynolds numbers, as appropriate for the neighborhood of a contact line. After some trial and error, values $g = O(0.1)$ were selected, which produce average velocities $u = O(0.1)$ and Reynolds numbers $<O(1)$.

Whereas an MD fluid at rest usually has a stable temperature, up to fluctuations, when motion occurs work is being done on the system and it tends to heat up. In principle the heat could escape through the walls, but in our case the wall molecules are so heavy that the wall is effectively an insulator. (In a somewhat more realistic treatment of a wall, one might assign the molecules a normal mass, but tether them to fixed positions with Hookes' Law springs[31].) We have followed either of two procedures with regard to temperature control. The first is to simply allow the fluid temperature to rise, which has the general drawback that the transport coefficients vary as the temperature does. In our runs the temperature can rise by as much as 65%, but the viscosity is known[32] to have a weak variation over the parameter range of interest. The second procedure is continued equilibration, in which all fluid velocities are rescaled at fixed intervals to maintain the mean kinetic energy; the drawback of this method is its unphysical nature. Fortuitously, in our work the statistical fluctuations are sufficiently large that we see the same results using either method.

To simulate Poiseuille flow, we let the system ''melt'' for 5000 time steps without acceleration, and then apply (11) with a non-zero value of g. The velocity in a given z bin $\{B_n : z_n < z < z_n + 1\}$ is computed as

$$\vec{u}_n = \frac{1}{N_n} \sum_{z \varepsilon Bn} \frac{d\vec{x_i}}{dt} , \qquad (6)$$

where $N_n$ is the number of molecules in bin n, and its x-component is fit to a parabolic profile

$$u(z) = (\rho g/2\mu) (z - z_1) (z_2 - z). \qquad (7)$$

The fit determines the viscosity $\mu$, as well as the locations $z_i$ where the velocity vanishes. In Fig. 4 we show the longitudinal velocity profiles at g = 0.1 for nine different sets of data points, corresponding to different intervals of 5000 time steps and different initial velocity distributions. The solid dots show the location and (zero) velocities of the two wall layers on each edge, and the solid line is an overall fit. The other components of u are consistent with zero. Although there is considerable spread in the data, each of the nine individual sets is well-fit by a parabola, in that a least-squares fit typically accounts for 99% of the variation. We see first that the fitted velocity vanishes at a distance approximately one layer spacing from each inner wall layer, so that the fluid molecules adjacent to the wall are on average

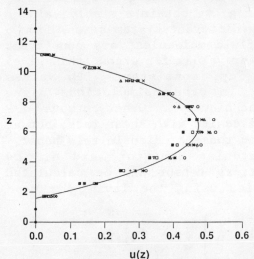

Fig. 4. Velocity profile in simulations of Poiseuille flow of a liquid at g = 0.1. The solid dots give the velocities and positions of the wall layers.

at rest. In other words, the no-slip condition has emerged naturally from a simulation based on a simple intermolecular interaction and walls with molecular roughness. We have repeated the calculation for different values of g, and we find global average values $z_1/\sigma$ = 1.6 ± 0.3 and $z_2/\sigma$ = 11.4 ± 0.3, with a slight tendency for $z_1$ to decrease and $z_2$ to increase with g. The fixed wall layers are located at $z/\sigma$ = 0, 0.86, 11.97, and 12.83, and so $z_1$ and $z_2$ are approximately the positions of the fluid layers of closest approach to the walls.

From an examination of the trajectories of the molecules, we have found no evidence for adhesion of a layer of fluid molecules to the wall for the potential we have used, but it is instructive to examine the effect of different wall interactions. We have examined two variant cases, wherein the attractive part of the LJ potential between fluid and wall is either doubled in strength or removed. In the former case there is a greater tendency for adhesion, and the location of the zeroes of the velocity profile moves deeper into the fluid. It is as though the trapped liquid layer has become part of the solid surface, and the location of no-slip occurs within $\sigma$ of the trapped layer. With no wall attraction, on the other hand, the velocity in the bin adjacent to the walls is rather larger, and the fitted locations of zero velocity move out into the walls a distance $O(\sigma)$. For all macroscopic purposes, however, in both of these variant cases the no-slip condition is still appropriate.

Returning to the viscosity, upon repeating the simulation and fit for different values of g, we obtain a stable value $\mu$ = 2.0 ± 0.3. Simulations using different system sizes, from 6 to 16 layers or 576 to 3200 fluid molecules, are consistent with this value. Using parameter values for argon, this gives a value of approximately 0.2cp, consistent with values obtained in experiment[33] or in other simulations using different non-equilibrium MD techniques[34]. Other authors[35,36] have observed a variation of viscosity with shear rate, but at even higher values of shear than those studied in this paper.

As a further test of Newtonian behavior, and of the viscosity computation, the stress tensor can be calculated from the molecular trajectories using the Irving-Kirkwood expression[37]

$$T = \sum_i m_i \left[ \frac{d\vec{x_i}}{dt} - \vec{u} \right] \cdot \left[ \frac{d\vec{x_i}}{dt} - \vec{u} \right] + \tfrac{1}{2} \sum_{ij} \vec{r}_{ij} \cdot \vec{F}_{ij}, \qquad (8)$$

where u (x) is the local average velocity and where $r_{ij}$ is the separation between molecules i and j, and $F_{ij}$ is the force

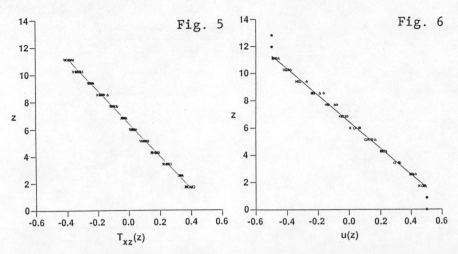

Fig. 5. Shear stress profile corresponding to Fig. 4.

Fig. 6. Velocity profile in Couette flow of a liquid, with the wall velocities and positions shown as solid dots.

between them. A binned version of (8) is computed as a running average during the calculation, with the result for the shear stress $T_{xz}$ shown in Fig. 5, for the same runs whose velocity is given in Fig. 4. The shear stress is linear, as expected from (7), and the other components of the tensor are in accord with continuum Poiseuille flow. Note that Newtonian behavior occurs despite the fact that in physical units the magnitude of the shear rate in this simulation is enormous, $O(10^{11}\ \mathrm{sec}^{-1})$. An independent calculation of the viscosity follows by dividing $T_{xz}$ by the derivative of the velocity profile, $\partial u/\partial z$, and again gives a value $\mu \approx 2.0$.

Yet another test of our conclusions is obtained by simulating plane Couette flow, where the two walls are given opposing constant velocities $u_w = \pm\ 0.5$ (with $g = 0$). The resulting velocity profile is shown in Fig. 6, where different runs and time intervals are superposed and the wall velocities are shown as solid circles. The profile is linear as expected and exhibits a no-slip boundary value, the shear stress is found to have a z-independent value $T_{xz} = 0.20 \pm 0.02$, and 2.0.

While no-slip has emerged from the simulation of liquids, a smooth transition to slip is found in the low-density, high-temperature ''fluid'' regime. In Fig. 7, we show sample velocity profiles obtained at $T = 2.5$ and a sequence of densities ranging from 0.8 to 0.08: the velocity at the wall systematically increases as the density decreases. We have

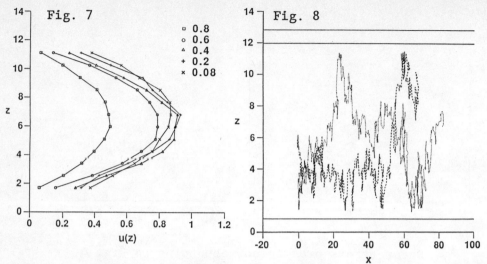

Fig. 7. Velocity profile in Poiseuille flow at T= 2.5 and decreasing densities as labeled.

Fig. 8. Molecular trajectories in Poiseuille flow; note the differing horizontal and vertical scales. The horizontal solid lines give the positions of the wall layers.

also computed the shear stress in these runs, and extracted a slip length using the Maxwell definition (1) and extrapolations of the fitted velocity and stress to the wall -- defined here as the center of the inner wall layer. A similar calculation was done for Couette flow at the same temperature and densities. We find that the slip length increases as the density drops, and further that the same slip length applies to two different flow configurations. This result supports the use of (1) as a genuine continuum boundary condition for gases, applicable to different flows.

Having examined the Eulerian properties of channel flow, we turn to the molecular motion. Figure 8 shows several individual molecular trajectories in Poiseuille flow; note the different horizontal and vertical scales. The typical motion might be described as Brownian, with a superimposed drift in the direction of the applied acceleration, and with some temporary localization near the fixed attracting wall molecules. As the fluid-wall interaction varies, the degree of wall localization changes in the obvious way. Note that there is no significant difference between molecules that were initially near the wall and those initially in mid-channel, in that the thermal motion eventually obliterates the initial bias. A last remark is that in a laboratory experiment with a realistic value of g and a macroscopic channel width, the relative amount of drift and wall localization is vastly smaller, and the molecular motion differs only slightly from Brownian.

## 5. Immiscible Fluids and Moving Contact Lines

We first introduce a mechanism for segregating immiscible fluids. A simple choice is to add an additional repulsive interaction between species,

$$V_{\alpha\beta}(r) = V_{LJ}(r) + (c_\alpha - c_\beta)^2 \cdot 4\varepsilon(r/\sigma)^{-6}, \tag{9}$$

where $c_\alpha$ is an adjustable ''pseudo-charge'' associated with species $\alpha$. The bulk interaction of pure fluid is thus unchanged, but we can vary the interaction between the two fluids A and B and with the wall W. The resulting intermolecular potential does not represent any specific pair of fluids that we are aware of, and more generally one could choose interactions specific to a realistic system of interest. However, this choice respects the physics of immiscibility and is effective in producing the desired separation of the fluids.

We study the motion of a contact line, by applying an acceleration to each fluid particle as before. A typical time sequence is shown in Fig. 9 at intervals of 8000 time steps after melting, for the choices $c_W = 0.1$ and $g = 0.1$. First we see that the two fluids maintain their phase separation even after two complete tranversals of the periodic channel. Secondly, note the distinct advancing and receding contact angles at the two A-B interfaces. The two angles differ both from the static angle and each other, and vary systematically with the velocity of the fluid. As in experiments,[4,5] with increasing velocity the advancing angle increases and the receding angle decreases. Unfortunately, we have found it difficult to quantify the latter variation because of the meniscus fluctuations at low velocity, and the limited range of higher velocities available in the low Reynolds number regime. Lastly, note the thin film of fluid A left behind along the walls. The film contains the fluid more strongly attracted to the wall, as one would expect, and has the additional consequence that the apparent receding contact angle effectively varies with time. When the bulk motion halts, on setting g to zero, the film begins to retract very slowly into the bulk, but apparently on a time scale long compared to the duration of the simulation. We have seen no evidence for a **precursor** film in the cases we have examined, but this absence could be due to such factors as insufficient

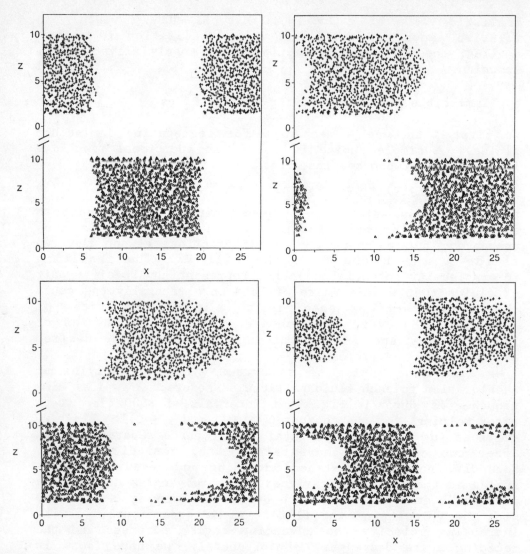

Fig. 9. Time sequence of two-fluid immiscible displacement and moving contact lines, as described in the text.

wall material to develop a reasonable van der Waals force, or simply not allowing enough time for a film to extend outwards from the bulk.

At high velocity the interface motion is fairly steady with a roughly constant velocity, for example $u_{int} \approx 0.2$ for the case shown in Fig. 9. In contrast, at very low acceleration and velocity, the bulk motion of the fluids is rather erratic and jumpy. For example at $g = 0.001$ the fluids remain more or less in place for 60000 time steps, then advance a distance

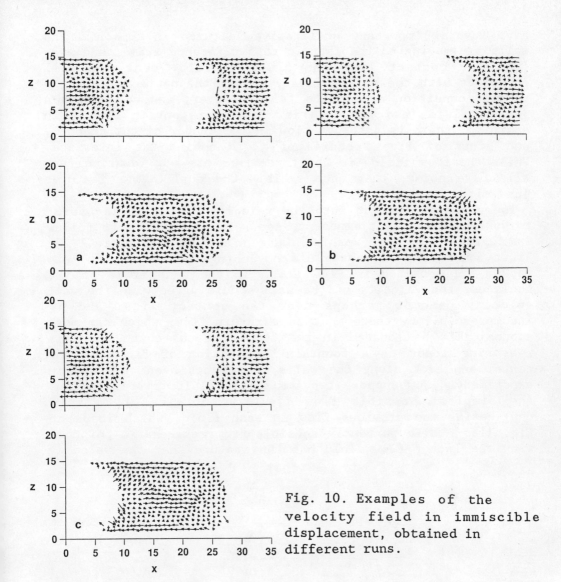

Fig. 10. Examples of the velocity field in immiscible displacement, obtained in different runs.

$O(\sigma)$ over the next 10000, then stop for (at least) another 30000 steps. Aside from being a clear manifestation of stick-slip motion, this behavior is suggestive of the phenomenon of contact angle hysteresis[4,5].

Next we consider the flow field in the moving contact line system. We have used a somewhat larger system with 8000 molecules (3200 per fluid and 800 per wall) and use a 40 x 20 array of **x-z** bins moving with the velocity of the interface, adjusting this velocity periodically. In Fig. 10a-c we show the average velocities of molecules occupying each bin over an interval of 12000 time steps. The three figures represent different initial distributions of molecular velocity, and

79

have several important qualitative features in common. The walls are moving to the left in this reference frame, and away from the contact line the fluid velocity near the wall coincides with the wall velocity. Near the contact lines the no-slip condition appears to fail, however, and we observe a jet of fluid into the interior from the contact line. In addition, there is a return flow in the center of the channel, and hints of closed eddies of fluid behind the interface. While the flow fields in the three runs are not identical and all quite noisy, they each exhibit the qualitative features just cited.

To gain some insight into the velocity field, it is useful to ask what one might expect to see. The discussion to follow is suggested by the work of Dussan and collaborators.[17,38] First consider a viscous fluid advancing through a slot into vacuum, illustrated in Fig. 11a. In the rest frame of the meniscus, the velocity of the wall should agree with the wall velocity, except perhaps near the contact line, and by incompressibility there must be a return flow, which can most economically be located in the center of the channel. This reasoning leads to a fountain-like motion of fluid up the center and back along the walls, which has been observed in experiment. Now suppose two immiscible fluids are in motion; the simplest possible flow field is obtained by simply duplicating the previous flow in each fluid, which leads to Fig. 11b. While perfectly sensible from a kinematic point of view the latter flow field has the feature that the velocity

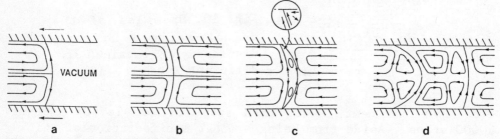

Fig. 11. Qualitative streamlines expected in various displacement situations, as described in the text.

must go to zero along the meniscus, because the velocities to either side of it are in opposite directions. This behavior can be ruled out by experiment, simply by placing some dye on the meniscus and observing that it moves outwards to the wall. The simplest means of allowing a non-zero velocity on the meniscus is to introduce a closed eddy in one fluid which reverses that fluid's velocity near the meniscus, as in Fig.

11c. The choice of which fluid has the closed eddy is not dictated by kinematics, but one might anticipate that it would occur as drawn in the wider of the two wedges, because this choice minimizes the bending of streamlines and would tend to lower the energy dissipation, in accord with both the detailed calculation of Huh and Scriven[3,39] and the laziness principle of Joseph et al.[40] Finally, if this reasoning is extended to the specific geometry of the present simulation, one obtains the flow field of Fig. 11d. While our simulation results in Fig. 10 do not exhibit this detailed streamline pattern, they are **consistent** with it, in that the reproducible features noted in the previous paragraph are all present.

Recently Thompson and Robbins[31] have studied the dependence of the interface shape and dynamic contact angle on the capillary number. Their results show that the no-slip boundary condition broke down within a couple of atomic spacings from the contact line. They suggest that the slip is associated with the breakdown of local hydrodynamic theory at atomic scales.

## 6. Conclusions

We have shown that molecular dynamics calculations can successfully reproduce the continuum properties of low Reynolds number flow near realistic solid boundaries, as well as answer questions that cannot be settled from a purely macroscopic viewpoint. In channel flows of a single fluid, where the interesting variation occurs only in one (spanwise) dimension, MD produces both the correct variation of the continuum velocity and stress fields and the Brownian and dispersive properties of the molecular motion. More complicated flows, involving two fluids or a two (or three) dimensional flow field, are difficult to fully resolve with system sizes currently acessible to MD, but anticipated advances in computing power should remove this restriction in the near future. Nonetheless, we have been able to draw significant qualitative conclusions for a particularly interesting two-fluid problem.

The no-slip boundary condition appears to be a natural property of a dense liquid interacting with a solid wall with molecular structure and interactions. There is some unavoidable ambiguity in assigning the precise location where the tangential velocity should vanish at a boundary, due to both a finite molecular size and to a quite reasonable variation with the details of the fluid-wall interaction. The fluid-wall interface is inherently ill-defined on the length scale of individual molecules, although any such imprecision is imperceptible at the continuum scale. Indeed, whenever such effects are important the continuum equations are insufficient.

The no-slip condition appears to fail at a moving contact line between immiscible fluids. One may describe this situation heuristically by saying that each fluid "tries" to stick to the wall, but the high shear rates required by the kinematics of a moving contact line overcomes the natural adherence. If no-slip is imposed, the shear rate tends to diverge, but the molecules are compliant enough to effect a compromise with some slip and high but finite shear.[41] A hint of the latter is seen in the MD simulations. Our results, with limited statistical resolution, do not suggest the precise form of the correct continuum boundary condition, but presumably further simulations of larger systems over longer times will settle this issue.

The more general conclusion is that even surprisingly small systems of several thousand molecules exhibit continuum behavior, and that MD is therefore a viable tool in continuum fluid mechanics. For most purposes MD involves far too much microscopic detail to be an efficient computational method, but in appropriately chosen problems it is a valuable addition to the repertoire of techniques. A particularly promising area is micro-hydrodynamics,[42] where one considers flows with length scales around a micron, plus or minus a decade or two, and in particular the derivation of effective transport properties of systems with non-trivial microscopic structure. A host of other problems may be studied with MD, ranging from the flow of more complicated (e.g., non-Newtonian or composite) fluids, to the effects of convection on reaction and phase change, to flows over irregular surfaces. We hope to address such problems soon.

Acknowledgements

We have benefitted from discussions with and the encouragement of Berni Alder, Stephen Garoff, Jacob Israelachvili, Mark Nelkin, John Ullo and Sidney Yip. We particularly thank Elizabeth Dussan for her continued interest in this work. We are grateful to the National Science Foundation and the Donors of the Petroleum Research Fund, administered by The American Chemical Society, for their support.

# References

1. S. Goldstein, ed., "Modern Developments in Fluid Dynamics" (Oxford University Press, Oxford, 1938), Appendix. We thank G. M. Homsy for bringing this reference to our attention.
2. G. K. Batchelor, "An Introduction to Fluid Dynamics" (Cambridge University Press, Cambridge, 1967).
3. C. Huh and L. E. Scriven, J. Coll. Int. Sci. 35, 85 (1971).
4. E. B. Dussan V., Ann. Rev. Fluid Mech. 11, 371 (1979).
5. P. G. De Gennes, Rev. Mod. Phys. 57, 827 (1985).
6. J. N. Israelachvili, P. M. McGuiggan and A. M. Homola, Science 240, 189 (1988).
7. D. D. Awschalom and J. Warnock, Phys. Rev. B 35, 6779 (1987).
8. M. P. Allen and D. J. Tildesley "Computer Simulation of Liquids" (Clarendon Press, Oxford, 1987).
9. G. Coccotti and W. G. Hoover, eds., "Molecular-Dynamics Simulation of Statistical Mechanics Systems" (North-Holland, Amsterdam, 1986).
10. J. M. Haile, "A Primer on the Computer Simulation of Atomic Fluids by Molecular Dynamics" (Clemson University, Clemson, SC, 1980).
11. E. Meiburg, Phys. Fluids 29, 3107 (1986); D. C. Rapoport, Phys. Rev. A 36, 3288 (1987).
12. M. Mareschal and E. Kestamont, Nature 329, 427 (1987); J. Stat. Phys. 48, 1187 (1987); D. C. Rapoport, Phys. Rev. Lett. 60, 2480 (1988).
13. B. J. Alder and T. E. Wainwright, Phys. Rev. A 1, 18 (1970); W. E. Alley and B. J. Alder, ibid. 27, 3158 (1983); W. E. Alley and B. J. Alder, Phys. Today 37 (1), 56 (1984).
14. J. Koplik, J. R. Banavar and J. F. Willemsen, Phys. Rev. Lett. 60, 1282 (1988); Phys. Fluids A1, 781 (1889).
15. J. C. Maxwell, Phil. Trans. Roy. Soc. 170, 231 (1867); see also R. Jackson, "Transport in Porous Catalysts" (Elsevier, Amsterdam, 1977).
16. G. N. Patterson, "Molecular Flow of Gases" (Wiley, New York, 1956); D. K. Bhattacharya and G. C. Lie, Phys. Rev. Lett. 62, 897 (1989).
17. E. B. Dussan V. and S. H. Davis, J. Fluid Mech. 65, 71 (1974).
18. See, e. g., G. E. P. Elliot and A. C. Riddiford, J. Coll. Int. Sci. 23, 389 (1967).
19. Y. Pomeau and A. Pumir, C. R Acad. Sci. 299, 909 (1984).
20. A. Nir, private communication (1988).

21. E. B. Dussan V., J. Fluid Mech. **77**, 665 (1976).
22. S. Richardson, J. Fluid Mech. **59**, 707 (1973); L. M. Hocking, J. Fluid Mech. **76**, 801 (1976).
23. K. M. Jansons, J. Fluid Mech. **154**, 1 (1985); Phys. Fluids **31**, 15 (1988).
24. J. N. Israelachvili and G. E. Adams, J. Chem. Soc. Faraday Trans. I **74**, 975 (1978).
25. J. N. Israelachvili, "Intermolecular and Surface Forces" (Academic, London, 1985).
26. F. F. Abraham, J. Chem. Phys. **68**, 3713 (1978); see also F. F. Abraham, Rept. Progr. Phys. **45**, 1113 (1982).
27. S. Toxvaerd, J. Chem. Phys. **74**, 1998 (1981).
28. J. J. Magda, M. Tirrell and H. T. Davis, J. Chem. Phys. **83**, 1888 (1985).
29. J. Q. Broughton and G. H. Gilmer, J. Chem. Phys., **64**, 5759 (1986), and earlier references therein.
30. J. Tallon, Phys. Rev. Lett. **57**, 1328 (1986).
31. P. A. Thompson and M. O. Robbins, Phys. Rev. Lett. **63**, 766 (1989).
32. W. T. Ashurst and W. G. Hoover, Phys. Rev. A **11**, 658 (1975).
33. S. A. Mikhailenko, B. G. Dudar and V. A. Schmidt, Sov. J. Low Temp. Phys. **1**, 109 (1975).
34. W. G. Hoover, D. J. Evans, R. B. Hickman, A. J. C. Ladd, W. T. Ashurst and B. Moran, Phys. Rev. A **22**, 1690 (1980).
35. D. M. Heyes, J. J. Kim, C. J. Montrose and T. A. Litovitz, J. Chem Phys. **73**, 3987 (1980).
36. J.-P. Ryckaert, A. Bellemans, G. Cicotti, and G. A. Paolini, Phys. Rev. Lett. **60**, 128 (1988).
37. J. H. Irving and J. G. Kirkwood, J. Chem. Phys. **18**, 817 (1950).
38. F. Y. Kafka and E. B. Dussan V., J. Fluid Mech. **95**, 539 (1979); E. B. Dussan V., AIChE J. **23**, 131 (1977).
39. Although their solution cannot be correct down to the contact line, it can represent an outer or matching solution when combined with a slip boundary condition in the inner region near the wall; see Ref. 38.
40. D. D. Joseph, K. Nguyen and G. S. Beavers, J. Fluid Mech. **141**, 319 (1984).
41. A contrasting situation occurs when a single fluid flows past a sharp corner, as discussed by K. Moffatt, J. Fluid Mech. **18**, 1 (1964).
42. G. K. Batchelor, in "Theoretical and Applied Mechanics", W. T. Koiter, ed. (North-Holland, Amsterdam, 1976).

# Computer Simulations for Polymer Dynamics

*K. Kremer*[1], *G.S. Grest*[2], *and B. Dünweg*[3]

[1]Institut für Festkörperforschung, Forschungszentrum Jülich,
  W-5170 Jülich, Fed. Rep. of Germany
[2]Corporate Research Science Laboratories, Exxon Research
  and Engineering Company, Annandale, NJ 08801, USA
[3]Institut für Physik, Universität Mainz, W-6500 Mainz, Fed. Rep. of Germany


Abstract. In this paper we review recent work on the dynamics of polymeric systems using computer simulation methods. For a two-dimensional polymer melt, we show that the chains segregate and the dynamics can be described very well by the Rouse model. This simulation was carried out using the bond fluctuation Monte Carlo method. For three-dimensional (3d) melts and for the study of hydrodynamic effects, we use a molecular dynamics simulation. For 3d melts our results strongly support the concept of reptation. A detailed comparison to experiment shows that we can predict the time and length scales for the onset of reptation for a variety of polymeric liquids. For a single chain, we find the expected hydrodynamic scaling for the mean square displacement and dynamic scattering function.

## I. Introduction

In the past few years computer simulations have proven to be a very powerful tool for investigating polymeric systems [1,2,3]. Since theoretical arguments are usually developed for ideal systems, *e.g.* a melt with no polydispersity, and experiments usually deal with rather complicated and nonideal systems, it is often difficult to adequately make comparisons between the two. Consequently, the applicability or approximations used in a particular theory are often very difficult to prove experimentally. It is the aim of many simulation studies to bridge the gap between these two extreme situations. In a computer simulation one can put in what one believes to be the essential ingredients, *e.g.* the chain connectivity, and then follow the monomers microscopically. A number of quantities can then be determined and compared to both experiments and theory. The major limitations of this approach are basically the speed and availability of the computer, the efficiency of the algorithms and the patience of the simulator. With recent advances in supercomputer hardware and software as well as the reduction in real cost for computer time, computer simulations have had a significant impact on our understanding of polymers in the past 3-5 years. We are sure that this trend will continue and simulations will become even more important in the future.

In order to use computer simulations effectively for macromolecular systems, it is necessary to consider very carefully the problem one would like to solve and whether it is appropriate for the computer resources available. One needs to first define the question under consideration very carefully and then choose the appropriate model and numerical technique. The available numerical methods range from a simple sampling procedure for static properties to a Monte Carlo (MC) or molecular dynamics (MD) simulation for dynamics. To demonstrate this aspect, in the present paper, we discuss three different topics, each one of which was studied with a different numerical technique, providing us with a set of methods which can be applied to a variety of other problems. The first problem we will discuss is the dynamics of a 2d polymer melt, which was carried out using a new, dynamic MC method referred to as the bond fluctuation method [4,5]. The second problem deals with the dynamics of 3d melts, in which a MD method is used, in which the monomers are coupled weakly to a heat bath [6,7]. The last problem deals with the dynamics of a dilute polymer in solution [8]. Since in this case we are interested in studying the effects of hydrodynamic interactions, we have to use a microcanonical MD simulation.

Before we discuss the details of these three systems, we would like to briefly review why it is important and in some cases essential to use coarse grained models for polymers and not attempt to construct detailed models on the atomic scale. Polymers are complicated molecules which contain thousands to hundreds of thousands of atoms. What is often referred to as a monomeric unit in fact contains roughly from 3 to 100 atoms depending on the complexity of the system. While one has to construct a precise polymeric model with more or less the complete chemistry of the molecule if one is interested in the local properties of a particular polymer, this is not necessary if one is interested in global properties. In order to understand general polymeric properties which are expected to be universal, it is better to use a model which contains only the essential interactions, such as the connectivity of the chain and excluded volume. For problems which involve time and length scales on the order of the chain relaxation times and chain extension, it is simply impossible to perform simulations with the full details of the chemistry, even assuming one can write reasonable classical interactions. For the simplest polymeric system, namely polyethylene (PE), simulations for the dynamics of a 3d melt would have taken of order 500,000 hours of CPU time on a single Cray XMP processor compared to about 1500 hours [9] for the same effective chainlengths as the study described in Sec. III. This large amount of time would be needed even for a model [10] for PE that treats the methyl $CH_2$ momoner as a single pseudoatom and weakens the C-C bond by about a factor of 7. For more complicated and/or interesting polymers, this time could easily increase by a factor of 10-100. An additional complication arises from the fact that the typical parameters used in the potential, except for simple interactions like the C-C spring constant, are not known very well. Even for PE there are a number of different interaction parameter sets in use [10,11]. Therefore the only procedure which we see that can be used today is to define the simplest model possible which contains the crucial interactions, carry out the simulations, and then map the model system onto chemical species for comparison to experiment.

In the rest of the paper, we will review some of our most important findings on the dynamic properties for three different polymeric systems. Results for the 2d polymeric melts will be presented in Sec. II. In Sec. III, we present our MD simulations for a 3d melt and a dilute chain in solution. In Sec. IV, we present a brief summary of our work and give a brief outlook on some of the new directions we feel computer simulations on polymers may take in the next years. These three methods have been used for a number of other problems which we do not have space to discuss here. For the bond fluctuation method these include problems on interdiffusion [12], electrophoresis [13], polymer glasses [14] and 2d ordering of stiff chains [15]. The MD approach has also been used to study many-arm star polymers [16,17], polymers attached to a repulsive surface [18] and gelation/percolation clusters and tethered membranes [19,20].

## II. Monte Carlo for Two-Dimensional Melts

Two-dimensional systems can be made by a variety of techniques. The method most often employed is to spread polymers from dilute solution containing a highly volatile solvent on top of a (liquid) surface. However, whether one obtains a system in which there are no overlaps is not clear. These overlaps may often survive for a very long time [21]. Besides being of interest experimentally, 2d polymers are also of considerable theoretical interest since the chain connectivity is supposed to affect the topology much more than in 3d. A 2d melt of linear chains is expected to segregate [22]. However, little is known about their dynamics.

The dynamics of polymer chains is typically discussed within the context of the Rouse model [22,23,24]. This model treats the Langevin dynamics of a simple bead-spring random walk with no excluded volume. The complicated intra- and interchain interactions are replaced by a bead friction $\varsigma$ and a heat bath $\vec{W}_i(t)$ acting on each monomer $i$. The heat bath and the friction are coupled via the fluctuation-dissipation theorem. For a polymer made of N bonds, the longest relaxation time $\tau_N = \varsigma <R^2(N)> /3\pi^2 k_B T$ for any dimension. Here $T$ is the temperature and $<R^2(N)>$ is the mean squared end-to-end distance

$$<R^2(N)> = <(\vec{r}_1 - \vec{r}_{N+1})^2>, \qquad (1)$$

where $\vec{r}_1$ and $\vec{r}_{N+1}$ are the first and last monomer on the chain. For an ideal random walk $<R^2(N)> = (ll_p)^2 N$, where $l$ is the bond length and $l_p$ is a persistence length which measures the chain stiffness. The longest relaxation time $\tau_N \sim N^2$. The diffusion constant $D$ is given by

$$D = k_B T/\varsigma N. \qquad (2)$$

Consequently the mean square displacement of the inner monomers $g_1(t)$ for times less than $\tau_N$ should be proportional to $t^{1/2}$. The question which then arises is how well does this theory work in two and three dimensions. In 2d, one would expect that, since the chains segregate and do not interpenetrate, the Rouse model should not be adequate to describe the internal dynamics of the chains. However, as shown below, the observed dynamical behavior compares surprisingly well to the Rouse model in 2d. This is not the case in 3d as shown in Sec. III, where the Rouse model works only for chains shorter than an entanglement length $N_e$.

Because of the strong influence on the excluded volume interaction, 2d melts are interesting systems to investigate. While both the bond fluctuation MC method on a lattice or a continuum MD simulation could have been used to study 2d melts, we chose the former method since it is a promising method, which can also be applied to systems where the discrete lattice structure is essential [14]. Note that standard kink-jump lattice MC methods can be used in 2d for our purposes. As seen in Fig. 1, the chains are confined to a square lattice. The elementary motions are given by jumps of a single monomer in

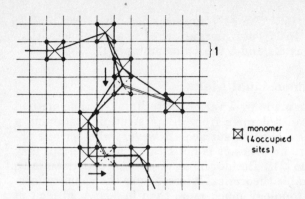

Figure 1: Illustration of the bond fluctuation method for a 2d branched polymer. Possible moves of the two monomers are indicated. Unlike other lattice models, this approach also allows for a dynamic MC simulation of branched objects and for d=2.

one of the four lattice directions. The bond length is restricted by the excluded volume and a maximal length $l \leq \sqrt{13}$ in 2d. The maximal length prevents the bonds from cutting each other. In 3d the bond length restriction is more complicated, however the generalization is straightforward. Because of the large internal flexibility of the chains in this model, it resembles a coarse grained flexible polymer chain more than a standard lattice model. The 2d algorithm can be easily vectorized by running many systems in parallel. In 3d, due to memory limitations, it is necessary to use a sublattice structure so that multiple moves can be made simultaneously in order to obtain good vectorization. On the Cray YMP at Jülich, we get up to $2.6 \times 10^6$ attempted monomer moves per second on one processor for the 3d system [25].

To follow the chain segregation, the chains initially had the shape of stretched internested U's with periodic boundary conditions. The systems spanned a density from $\rho = 0.20$ to $0.80$ with chain lengths from N=20 to 100. Figure 2 shows two typical snapshots of a typical configuration for N=20 and 100 at two dif-

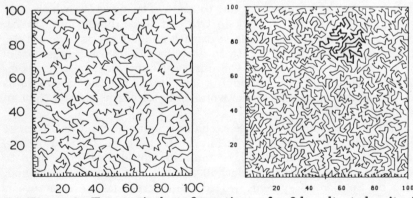

Figure 2: Two typical configurations of a 2d melt at density for N=20 at $\rho = 0.40$ and for N=100 at $\rho = 0.80$. The N=100 structure is almost equilibrated and the chains are segregated.

ferent densities. The 100 sample is probably not fully equilibrated. The chains segregate and become compact spherical objects, with size fluctuations which are of order their own diameter. The mean square radius of gyration $< R_G^2(N) >$ and $< R^2(N) >$ display the expected scaling behavior,

$$< R^2(N) > \sim < R_G^2(N) > \sim N. \tag{3}$$

The average number of neighbor chains in contact with a given chain is 6. Thus a 2d polymer melt can be viewed as a liquid of very soft disks.

Whether the dynamics of this fluid of segregated chains resemble the Rouse model is a question that is perfect for computer simulations to answer. From the soft sphere fluid picture one expects that motion will occur via shape fluctuations with a diffusion constant proportional to $N^{-1}$. This simple picture is confirmed by the data. The relaxation time of the autocorrelation function of the mean square radius of gyration is found to scale as $N^2$. This relaxation time compares very well to the diffusion time, Eq. (2). This result agrees with the simple Rouse picture. For the motion of the monomers the situation is more complicated since they have to move along each other and cannot jump over one another. Figure 3 shows the results for $g_1(t)$ for the inner and outer monomers for $\rho = 0.80$ and N=50. The data are in very good agreement with the Rouse model, namely $t^{1/2}$. To understand why this is true even though the Rouse model does not take into account the topology of the chains, we also analyzed the normal modes of the chain. The relaxation time $\tau_{p,N}$ for the $p^{th}$ normal mode of a chain should scale as [24]

$$\tau_{p,N} = \tau_{1,N}/p^2. \tag{4}$$

Surprisingly, again one finds [5] that the Rouse scaling works very well. However, we did find deviations in the prefactor, which can be mapped onto the monomeric

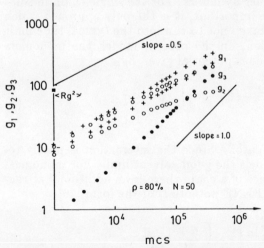

Figure 3: Mean square displacements for inner and outer monomers of a chain of N=50 at $\rho = 0.80$. $g_1(t)$ corresponds to the motion of the monomers in space, $g_2(t)$ is in the reference frame of the center of mass of the chains while $g_3(t)$ describes the motion of the center of mass itself.

friction coefficient $\varsigma$. There is a significant distinction between the simple Rouse model and the real 2d melts. Since both the diffusion constant and relaxation time for the modes contain $\varsigma$, the two should give the same result for $\varsigma$ for the theory to be consistent. However, we find that, depending on the density, $\varsigma$ determined from the normal modes is 3-5 times larger than that calculated from the diffusion constant. The discrepancy increases as the density increases. Thus the internal reorganization is slower than the diffusion via shape fluctuations. A similar effect is expected to occur in star polymers [17] or more generally in branched polymers. From the difference in the two friction constants, we can conclude that the Rouse-like appearance of the chain dynamics is rather accidental. It is more likely due to the coincidence of the space dimension and the fractal dimension of a random walk, on which the Rouse model is based. From this point of view, it would be very interesting to study single collapsed chains in higher dimensions.

## III. Molecular Dynamics Simulations

As discussed in the introduction, one of the most important considerations in simulating polymers is to choose the most efficient model for the question under investigation. For the dynamics of polymer chains in a melt one important question is: How does the chain topology influence the dynamics? Our experience from experiments and earlier simulations shows that in order to study this effect one has to work at relatively high density. However, at high density standard, stochastic MC algorithms encounter problems and thus are restricted to low densities [2]. The bond fluctuation model is one method which overcomes many of the difficulties of standard lattice MC. It allows for efficient simulations of dense systems at relatively high density with the advantages of discrete space. However, the algorithm is stochastic. Another approach is to find a method in which the motion is dominated by the local interactions. MD is such a method and is appropriate for studying polymer dynamics. For the single chain in solution, where hydrodynamic effects are important, it is the only approach which can be used. However, care must be taken not to run into the typical time limit problems discussed in the introduction. In the present model, the monomers interact through a shifted Lennard-Jones potential given by

$$U^0(r) = \begin{cases} 4\epsilon\left[\left(\frac{\sigma}{r}\right)^{12} - \left(\frac{\sigma}{r}\right)^6 + \frac{1}{4}\right] & \text{if } r \leq r_c; \\ 0 & r > r_c, \end{cases} \quad (5)$$

where $r_c = 2^{1/6}\sigma$ is the interaction cutoff. Since the interaction is purely repulsive, for an isolated chain this models the good solvent limit. For monomers which are connected along the sequence of a chain there is an additional attractive interaction potential (called the FENE potential) of the form [26]

$$U^{bond}(r) = \begin{cases} -0.5kR_0^2 \ln\left[1 - \left(\frac{r}{R_0}\right)^2\right] & \text{if } r \leq R_0; \\ \infty & r > R_0, \end{cases} \quad (6)$$

modelling an anharmonic spring. The parameters are then adjusted according to the problem being studied.

## A. Polymer Melts

For the 3d polymer melts the parameters were set to $k = 30\epsilon/\sigma^2$, $R_0 = 1.5\sigma$ and $k_BT = 1.0\epsilon$. The system simulated covered the range from $N = 5$ monomers per chain up to $N = 400$. Typically our samples contained $M = 20$ chains (for N=400, there were only 10 chains). We also investigated one system of $M = 100$, $N = 200$ to test finite size effects. The density was $0.85\sigma^{-3}$. For the equilibrated structures our chains had a persistence length of $ll_p = 1.3\sigma$. Thus the longest chain studied contained almost 300 persistence lengths.

Here we want to consider the dynamics of 3d polymer melts, which has been discussed in the literature for more than 20 years [24]. For chains shorter than a characteristic length $N_e$, the motion is well described by the Rouse model. This is somewhat surprising since the Rouse model accounts for all the complicated interactions between individual monomers only via a heat bath and friction with the background. When the chainlength significantly exceeds $N_e$, the motion slows down dramatically. The diffusion constant D is found to be proportional to $N^{-2}$ and the longest relaxation time $\tau_R \sim N^{3.4}$. A variety of concepts have been proposed to explain this behavior. The most widely used model is the repation model of Edwards [27] and de Gennes [28]. The idea is that the topological constraints of the chains, on a coarse grained length scale, which exceeds the diameter of a coil of $N_e$ monomers, cause the monomers to move only along their own contour. The chain is supposed to move along a tube of diameter $d_T = < R^2(N_e) >^{1/2}$. The diffusion along the tube results in a diffusion constant $D \sim N^{-2}$ as found in experiment. However, the longest relaxation time of the chains is $\tau_R \sim N^3$ instead of the experimental result $N^{3.4}$ from viscosity measurements. There are many different attempts to explain this discrepancy between 3 and 3.4 as a crossover phenomenon [24] however some take it as evidence that the entire reptation concept is incorrect [29]. Therefore it is important to test this from a microscopic point. In such a situation computer simulations can be an important tool to bridge the gap between the theoretical concepts and current experiments. In order to shed some light on these problem s extremely long runs are needed. For the present investigation we solved Newton's equation of motion, but also coupled each monomer weakly to a heat bath and a frictional background. This provides us with a stable algorithm which simulates a canonical ensemble [7]. Denoting the total potential of monomer $i$ by $U_i$, the equation of motion for monomer $i$ is given by

$$m\frac{d^2\vec{r_i}}{dt^2} = -\vec{\nabla}U_i - m\Gamma\frac{d\vec{r_i}}{dt} + \vec{W_i}(t). \tag{7}$$

Here $m$ is the monomer mass, and $\Gamma$ is the bead friction which acts to couple the monomers to the heat bath. $\vec{W_i}(t)$ describes the random force acting on each bead. It can be written as a Gaussian white noise with

$$\langle \vec{W_i}(t) \cdot \vec{W_j}(t')\rangle = \delta_{ij}\delta(t-t')6k_BT\Gamma. \tag{8}$$

We chose $\Gamma = 0.5\tau^{-1}$ and $k_BT = 1.0\epsilon$. Here $\tau = \sigma(m/\epsilon)^{1/2}$. The equations of motion are then solved using either a third- or fifth-order predictor-corrector algorithm [30] with a time step $\Delta t = 0.006\tau$. The program was vectorized following the method discussed in Ref. 31. Further details of the method can

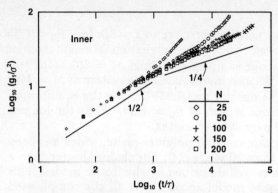

Figure 4: Mean square displacement $g_1(t)$ verus $t/\tau$ averaged over the inner 5 monomers for five values of N.

be found elsewhere [6,7]. In the course of this investigation many different quantities were analyzed. Here we only discuss the motion of the monomers. Following the reptation picture the monomers follow a Rouse relaxation until they feel the constraints of the tube. Then the Rouse relaxation is confined to the tube, which is the coarse grained contour of the chain which also has a random walk structure. Consequently the initial $t^{1/2}$ power law for the monomer motion crosses over to a $t^{1/4}$ power law for distances greater than the tube diameter and times longer than the Rouse relaxation time of a subchain of length $N_e$. After the Rouse time of the whole chain ($\sim N^2$) the chain still has only moved a distance proportional to $N^{1/2}$ along the tube. The subsequent diffusion along the tube produces a second $t^{1/2}$ regime in the displacement in space. Only after the chain has left the original tube, $t > \tau_R \sim N^3$, can one observe the free diffusion. Fig. 4 shows the mean square displacements of the inner monomers. One clearly observes a strong slowing down in the motion, which turns out to be independent of chain length. This indicates a unique length scale. The slope approaches the $t^{1/4}$ power law, though we actually measure a value of 0.28. From this and a detailed normal mode analysis we find that $\tau_e = 1800\tau$ and $N_e = 35$ for our model. Combining all the different data, we find strong evidence for reptation compared to other competing models, although our chains only contain a few entanglements.

Experimentally there has been a long-standing controversy about whether one can observe reptation on a microscopical length and time scale. Several neutron spin echo experiments [32,33] produced contradictory results. To understand this problem we have to map our computer polymers to chemical polymers. However, this cannot be done by mapping the persistence lengths. The characteristic length in the problem is the entanglement length $N_e$. The characteristic time is the relaxation time $\tau_e$ of a subchain of length $N_e$. To check this mapping, we compare the diffusion constants for several different systems. If we normalize the diffusion constant $D(N)$ by the extrapolated Rouse diffusion constant $D_{Rouse}(N)$ and plot this versus $N/N_e$ (or $M/M_e$, respectively) we should obtain a universal curve. Fig. 5 shows such a comparison for our data and for polyethylene (PE) [34] and polystyrene (PS) [35]. It shows a very good agreement between PE and the MD model while there are deviations from PS.

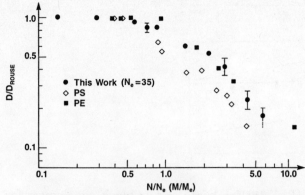

Figure 5: Diffusion data for PS (Ref. 35) and PE(Ref. 34) compared to the simulation results [6].

Taking the error bars for the experimental diffusion constants, this deviation is significant. Therefore this raises some questions about the validity of universality for dynamics in this crossover regime. Since PS has a small side group and PE does not, just as our model, we believe that the coincidence with PE is more significant than the deviations from PS. This mapping now enables us to predict time and length scales for the various chemical systems. For polymers whose monomeric units do not contain large side groups we expect this mapping to be quite accurate. Comparing the entanglement molecular weight gives a direct estimate of the tube diameter. From the monomeric friction coefficient $\varsigma$ or the short chain diffusion constant compared to the results from the simulation, we get the time mapping. This fixes the time $\tau_e$, the relaxation time of a chain of length $N_e$. The time $\tau_e$ is the time when the deviations from the Rouse behavior become significant. Using the above mapping, we find that for PDMS at 373K, $\tau_e = 10^{-7}s$, for PTHF at 418K, $\tau_e = 3.2 \times 10^{-9}s$ while for PEP at 500K, $\tau_e = 1.0 \times 10^{-8}s$. The estimated times for PDMS and PTHF for the temperatures of the experiments show that the neutron spin echo experiments on PTHF should see a deviation from Rouse. For PDMS this is not the case, since the times are beyond the resolution of the experiment, which was around $10^{-8}$ seconds. This resolves the contradictory experimental findings of the early neutron spin echo work [32,33].

Very recently Richter *et al*. [36] have investigated very well characterized polyethylene-propylene (PEP) samples. They used an improved spin echo machine with a resolution of $4 \times 10^{-8}$ seconds. For PEP we predict a crossover time of $1.0 \times 10^{-8}s$, which is within the resolution of the new instrument. For times longer than $\tau_e$ but shorter than $\tau_R$ (the actual longest time depends on the wave vector $q$ [37]) the single chain scattering function $S(q,t)$ should deviate from the simple Rouse decay and approach a constant of $S(q,t \gg \tau_e) = 1 - q^2 d_T^2/36$. This is because for these times the scattering function only measures a smeared out density in the tube. If we extrapolate from this plateau value to short times, the crossing time with the Rouse relaxation curve should be $\tau_e$. Fig. 6 shows this for two extreme scattering vectors for PEP. The agreement with the data is excellent, displaying the overall consistency of the numerical and experimental results combined with the reptation concept.

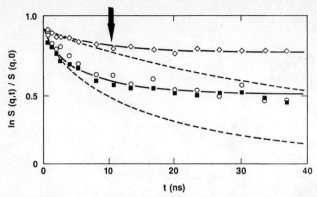

Figure 6: S(q,t) for PEP (Ref. 36) for homo and triblock copolymers for $q = 0.078\text{Å}^{-1}$ (upper curve) and $q = 0.116\text{Å}^{-1}$ (lower curve). Our result $\tau_e = 1.0 \times 10^{-8}$ is indicated by the arrow. The dashed lines indicate the expected intermediate scattering function for Rouse relaxation.

## B. Polymer in Solution

For the single polymer in solution the situation is completely different. The dynamics of the single chain is usually accounted for by the Zimm model [38,24]. This is a generalization of the Rouse model, which takes hydrodynamic interactions into account. The equation of motion (in the Ito interpretation) for monomer $i$ is given by

$$\frac{d\vec{r}_i(t)}{dt} = \sum_j \underline{\underline{H}}_{ij}(-\frac{\partial U}{\partial \vec{r}_j}) + \vec{W}_i(t) \tag{9a}$$

with a generalized fluctuation-dissipation theorem

$$<W_i^\alpha(t)W_j^\beta(t')> = 2k_B T H_{ij}^{\alpha\beta}\delta(t-t'), \tag{9b}$$

where $U$ is the potential energy between monomers. In the Rouse and Zimm models this is just the interaction from the harmonic spring which connects two monomers. $\underline{\underline{H}}$ is the Oseen tensor

$$\underline{\underline{H}}_{ij} = \frac{1}{8\pi\eta_s|\vec{r}_i - \vec{r}_j|}(\underline{1} + \hat{r}_{ij}\hat{r}_{ij}) \quad for\ i \neq j \tag{10}$$

and $\underline{\underline{H}}_{ii} = \varsigma^{-1}\underline{1}$. Here $\eta_s$ the solvent viscosity, $\hat{r}_{ij}$ is the unit vector of $\vec{r}_i - \vec{r}_j$ and $\hat{r}_{ij}\hat{r}_{ij}$ is a tensor product. For $\underline{\underline{H}}_{ij} = \varsigma^{-1}\delta_{ij}\underline{1}$, Eq. (9) is just the Rouse equation of motion. In most investigations a simplified form of Eq. (9) is used, where one only uses the average $<1/|\vec{r}_i - \vec{r}_j|>$. This is known as the preaveraged Oseen tensor. For the diffusion constant Eq. (9,10) yield

$$D = k_B T / 6\pi\eta R_H + k_B T / \varsigma N, \tag{11}$$

where $R_H$ is the hydrodynamic radius of the chain

$$R_H^{-1} = \frac{1}{N^2} \sum_{i,j; i \neq j} <1/|\vec{r}_i - \vec{r}_j|>. \qquad (12)$$

For long chains $<R_H^{-1}> \sim <R^2(N)>^{1/2}$, however the crossover to the asymptotic regime is rather slow [39,40]. Again we see that there are a number of assumptions. First, the above approximation requires that the monomers under consideration have to be far apart, since the Oseen tensor is only the leading term in a $\frac{1}{r}$ - multipole expansion of the hydrodynamic interaction. Second, as in the Rouse model, the solvent is taken as an incompressible homogeneous background. On the length scales of a few monomers this is certainly not correct, since many typical organic solvents, e.g. cyclohexane, are of comparable size to the monomers themselves. Typical experiments, such as dynamic light scattering, can measure $S(q,t)$ and find the expected power law (see below). However $R_H$ can only be measured via the diffusion constant itself.

In order to shed some more light on the intrinsic properties of a single chain in solution, we performed a MD simulation. One could also try to directly simulate Eq. (9,10) via Brownian dynamics, however, we found that the algorithms become unstable because of problems arising from monomers which approach each other too strongly [8]: The diffusion tensor then gets negative eigenvalues which corresponds to a completely unphysical negative dissipation.

In order to perform a MD simulation we explicitly take the solvent into account and use a microcanonical simulation. When the particles are coupled to a heat bath, one can show that the hydrodynamic interactions are screened on a length scale [41]

$$l_0 = \sqrt{\frac{\eta_s}{\rho \Gamma}}, \qquad (13)$$

where $\rho$ is the particle number density. We use the same model as for the melt study, however with different parameters for the anharmonic spring, Eq. (6). The chain was immersed in a good solvent of monomers at a density of $0.864\sigma^{-3}$. The largest system contained a total of 8000 particles including a single chain of length N=60. The parameters were $k_B T = 1.2\epsilon$, $k = 7\epsilon/\sigma^2$, $R_0 = 2\sigma$. The mass of the chain monomers was twice the mass of the solvent particles. In order to generate many different independent starting states the particles were coupled to a heat bath just as for the melts. We then turned off the coupling to the heat bath so that we had a microcanonical ensemble. The timestep was $\Delta t = 0.004\tau$. The typical performance on a single Cray YMP processor was about $3 \times 10^5$ particle moves per second. The statics shows the expected behavior of a good solvent chain with an exponent $\nu = 0.580 \pm 0.01$, for the chain lengths between $N = 30$ and $N = 60$.

In the Zimm model [24,38] the diffusion constant can be visualized by moving a rigid object of diameter $R_H$ through the solvent. From Eq. (11) one then finds that the longest relaxation time of the chain $\tau_Z \sim N^{3\nu}$. For the motion of the monomers one finds

$$g_1(t) \sim t^{2/3}, \qquad (14)$$

independent of $\nu$. We find very good agreement with Eq. (14) with an exponent in the range 0.67 to 0.70. Unfortunately $g_1(t)$ cannot be measured directly experimentally. What is measured is the time dependent structure function $S(q,t)$

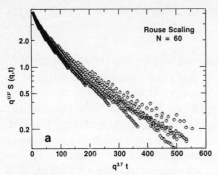
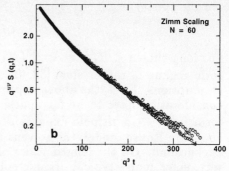

Figure 7: Dynamic scattering function of a polymer chain in a solvent with (a) Rouse scaling of the time, $q^{3.7}t$ and (b) Zimm scaling of the time $q^3 t$.

of a single chain. It should show a scaling with a scaling argument $q^3 t$ instead of $q^{3.7}t$ (Rouse), which comes from the different motion of the monomers with and without hydrodynamic interaction, respectively. Fig. 7 shows a comparison of the good solvent Rouse scaling compared to Zimm scaling. We find very good agreement with the Zimm model. We also found that the measured diffusion constant is much smaller than the Kirkwood prediction, Eq. (11), if one takes into account only the first term and evaluates $\eta_s$ by a Green-Kubo integration and $R_H$ by the definition (12). This difference is mainly due to the fact that the theory does not take into account the hydrodynamic interaction of the chain with its periodic images. However, this effect may be incorporated into the theory by Ewald sums [42], and, in our case, increases the effective hydrodynamic radius by roughly a factor of two. Taking this correction into account, we also find good agreement with the Zimm model for $D$.

## IV. Conclusion

This paper briefly reviewed some recent work on the dynamics of polymer systems. We showed that simulations can sucessfully bridge the gap between experiment and theory. We have been able to test several basic ideas and to predict explicitly the time and length scales for experiments, where the theory only gives general relations. For 2d systems no analytical theory exists up to now which can be used to rationalize our findings in detail. Experimentally it is very difficult to investigate strictly 2d systems, though a great effort is being made at several institutions. For the 3d melts we were able to show that already in the crossover regime from Rouse to highly entangled systems the reptation concept provides a good description of the data. Using an analysis of the Rouse normal modes of the chains some models for the dynamics can be excluded. With the mapping onto experimental systems, we were able to resolve a longstanding controversy between different neutron spin echo experiments. For the case of the chain in the solvent, this was the first time simulations were able to study the effect of hydrodynamic interactions. Based on this initial study a detailed and thorough investigation of these problems should be continued.

Based on the experience of the above discussed examples, we think we are in a good position to investigate many, more complicated important problems in polymer physics. The more complicated the systems under investigation the more difficult it will be both to perform a well-controlled experiment and a thorough quantitative theory. We think computer simulations will become a more and more important method and in many cases will provide the first detailed and well-controlled results for a number of systems. Such systems include polymer networks, rubbers, copolymers, polyelectrolytes, and micelles to name a few. For networks and rubber, simulations should provide the first direct microscopic insight into the significance of trapped entanglements. In contrast to experimental systems, the number of crosslinks and the strand length between crosslinks can be controlled. For copolymers, simulations are just at their infancy [43], but significant progress should be expected in the next few years. In the case of polyelectrolytes, analytical theory has trouble dealing with the electrostatic interactions properly beyond the simple Debye-Hückel approximation. Here numerical studies which can take the polyions and counterions explicitly into account will be very useful to sort out what is happening in a number of experimental systems. From this point of view, with a decrease in computing costs, we think computer simulations for such complex structures are just beginning to play an important role.

## Acknowledgements

This research has benefitted from discussions and collaboration with K. Binder, I. Carmesin, P. A. Pincus and H. P. Wittmann. GSG and KK acknowledge support from NATO travel grant 86/680. BD is supported by a grant from the Deutsche Forschungsgemeinschaft, DFG. Most of the calculations presented in this paper were made possible by a generous grant of computer time from the German supercomputer center HLRZ, Jülich.

## REFERENCES

1. A. Baumgärtner, *Ann. Rev. Phys. Chem.* **35**, 419 (1984).
2. K. Kremer and K. Binder, *Comp. Phys. Rept.* **7**, 261 (1989).
3. K. Binder, in *Molecular Level Calculations of the Structure and Properties of Non-Crystalline Polymers*, edited by J. Biscerano (Dekker, New York, 1989).
4. I. Carmesin and K. Kremer, *Macromolecules* **21**, 2819 (1989).
5. I. Carmesin and K. Kremer, *J. Phys. (Paris)*, 1990.
6. K. Kremer, G. S. Grest, and I. Carmesin, *Phys. Rev. Lett.* **61**, 566 (1988); K. Kremer and G. S. Grest, *J. Chem. Phys.*, 92, 5057, 1990.
7. G. S. Grest and K. Kremer, *Phys. Rev. A* **33**, 3628 (1986).
8. B. Dünweg and K. Kremer, (to be published, 1990).
9. K. Kremer and G. S. Grest, in *Computer Simulations of Polymers*, edited by R. J. Roe (to be published).
10. D. Rigby and R. J. Roe, *J. Chem. Phys.* **87**, 7285 (1987); **89**, 5280 (1988).
11. J. H. R. Clarke and D. Brown, *Mol. Sim.* **3**, 27 (1989).
12. W. Jilge, I. Carmesin, K. Kremer, and K. Binder, (to be published, 1990).
13. J. Batoulis, N. Pistoor, K. Kremer, and H. L. Frisch, *Electrophoresis* **10**, 442 (1989).

14. H. P. Wittmann, K. Kremer, and K. Binder (to be published, 1990).
15. A. Lopez Rodriguez, H. P. Wittmann, and K. Binder (to be published, 1990).
16. G. S. Grest, K. Kremer, and T. A. Witten, *Macromolecules* **20**, 1376 (1987).
17. G. S. Grest, K. Kremer, S. T. Milner, and T. A. Witten, *Macromolecules* **22**, 1904 (1989).
18. M. Murat and G. S. Grest, *Phys. Rev. Lett.* **63**, 1074 (1989).
19. F. Abraham, W. E. Rudge and M. Plischke, *Phys. Rev. Lett.* **62**, 1757 (1989).
20. G. S. Grest and M. Murat, *J. Phys. (Paris)*, 1990.
21. K. Kremer, *J. Phys. (Paris)* **47**, 1269 (1986).
22. P. G. de Gennes, *Scaling Concepts in Polymer Physics* (Cornell Univ. Press, Ithaca, NY, 1979).
23. P. E. Rouse, *J. Chem. Phys.* **21**, 1273 (1953).
24. M. Doi and S. F. Edwards, *The Theory of Polymer Dynamics* (Clarendon Press, Oxford, 1986).
25. H. P. Wittmann and K. Kremer (to be published, 1990).
26. R. B. Bird, R. C. Armstrong, and O. Hassager, *Dynamics of Polymeric Liquids* (Wiley, New York, 1977), Vol. 1.
27. S. F. Edwards, *Proc. Phys. Soc.* **92**, 9 (1967).
28. P. G. de Gennes, *J. Chem. Phys.* **55**, 572 (1971).
29. For a review see T. P. Lodge, N. A. Rotstein and S. Prager, Adv. Chem. Phys. (to be published, 1990).
30. C. W. Gear, *Numerical Initial Value Problems in Ordinary Differential Equations*, Prentice-Hall, Englewood Cliffs, NJ (1971).
31. G. S. Grest, B. Dünweg, and K. Kremer, *Comp. Phys. Comm.* **55**, 269 (1989).
32. J. S. Higgins and J. E. Roots, *J. Chem. Soc. Faraday Trans. 2* **81**, 757 (1985).
33. D. Richter, A. Baumgärtner, K. Binder, B. Ewen, and J. B. Hayter, *Phys. Rev. Lett.* **47**, 109 (1981).
34. D. S. Pearson, G. Verstrate, E. von Meerwall, and F. C. Schilling, *Macromolecules* **20**, 1133 (1987).
35. M. Antonietti, H. K. Foelsch, and H. Sillescu, *Makrom. Chem.* **188**, 2317 (1987).
36. D. Richter, B. Farago, C. Lartique, L. J. Fetters, J. S. Huang, and B. Ewen, (to be published, 1990).
37. K. Kremer and K. Binder, *J. Chem. Phys.* **81**, 6381 (1984).
38. B. Zimm, *J. Chem. Phys.* **24**, 269 (1956).
39. J. des Cloizeaux, *J. Phys. Lett. (Paris)* **39**, L-151 (1978).
40. J. Batoulis and K. Kremer, *Macromolecules* **22**, 4277 (1989).
41. B. Dünweg, in preparation, 1990.
42. C. W. J. Beenakker, *J. Chem. Phys.* **85**, 1581 (1986).
43. B. Minchau, B. Dünweg, K. Binder (to appear in Polymer Communications, 1990).

# Computer Simulation Studies of Phase Transitions in Two-Dimensional Systems of Molecules with Internal Degrees of Freedom

*O.G. Mouritsen[1], D.P. Fraser[2], J. Hjort Ipsen[1], K. Jørgensen[1,3], and M.J. Zuckermann[2]*

[1]Department of Structural Properties of Materials, The Technical University of Denmark, Building 307, DK-2800 Lyngby, Denmark
[2]Department of Physics, McGill University, Montreal, PQ, H3A 2T8, Canada
[3]The Royal Danish School of Pharmacy, Universitetsparken 2, DK-2100 Copenhagen Ø, Denmark

This paper presents a brief overview including recent results obtained from simulation studies of models of pseudo-two-dimensional systems of molecules with translational (crystalline) as well as internal (conformational) degrees of freedom. The models, which include both lattice-gas Potts models and models of hard discs with varying sizes, have a general sphere of applicability. The models are here being used to describe the phase transitions between the condensed phases in lipid monolayers or lipid bilayers. The simulation results reveal an intricate interplay between ordering processes governed by the two different degrees of freedom.

## 1. Introduction

Condensed molecular systems which possess more than one type of degree of freedom, for example both translational and internal degrees of freedom, are known to display a particularly rich phase behavior. Striking examples include three-dimensional nematic systems and two-dimensional molecular overlayers adsorbed on solid substrates. In these systems, the coupling between the translational and internal variables is responsible for a variety of phases, for example orientationally ordered liquids and glasses, orientationally ordered solids, orientationally disordered liquids, orientationally disordered solids, and orientational glasses.

There exists a particularly rich and interesting class of physical realizations of two-dimensional models of molecules with translational as well as internal degrees of freedom. These are lipid membranes [1,2]. Lipid membranes are aqueous lamellar aggregates of lipid molecules and they do not require a solid support for their stability. Lipid molecules, like certain soaps and detergents, are amphiphilic molecules with flexible hydrocarbon chains. When immersed in water these molecules, due to the hydrophobic effect, form stable condensed *bilayers* of a two-dimensional planar lamellar symmetry (Fig. 1(a)). When spread on an air/water interface, and subject to an external lateral pressure, the molecules form *monolayers* in condensed phases (Fig. 1(b)). Lipid membranes exhibit a number of phase transitions of which we shall only consider the so-called bilayer main transition (or monolayer liquid-extended – liquid-expanded transition). At the main transition, the bilayer goes from a low-temperature acyl-chain conformationally ordered phase (the gel phase) to a high-temperature acyl-chain conformationally disordered phase (the fluid phase). Hence this transition is usually thought of as a transition which predominantly involves the internal degrees of freedom, that is, the hydrocarbon flexibility, and is consequently referred to as a chain-melting transition. There are several reasons for this. Firstly, experimentalists have been preoccupied with measurements of acyl-chain order parameters. Secondly, the main transition, which is of first order,

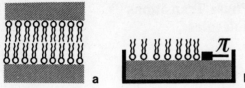

**Fig. 1.** Schematic cross-sections of a lipid bilayer membrane (a) and a lipid monolayer membrane (b).

typically involves a transition entropy of the order of $15k_B$/molecule in contrast to the $0.5k_B$ associated with the spatial disordering estimated for a two-dimensional melting process [3]. Both of these reasons underlie the fact that the major part of the theoretical modelling so far of lipid membrane phase transitions is concerned with the acyl-chain degrees of freedom, the single-chain statistics, and the interaction between different chain-conformational states [4–6]. However, it is imperative for a deeper understanding of the lipid membane as a two-dimensional system with different types of internal structure and global symmetries to extend the modelling to include internal as well as lateral, translational degrees of freedom and their mutual coupling. This is particularly true for the modelling of impurity effects since membrane-bound impurities, such as proteins and cholesterol, often couple quite differently to the two degrees of freedom leading to the peculiar phase behavior of many-component membrane systems which is so essential for biological viability [7]. Furthermore, although the conformational variables dominate the equilibrium properties of the main transition of pure lipid membranes, non-equilibrium features are likely to be strongly influenced by the crystalline (or glassy) nature of the solid phases, which invariably involve the translational variables.

The difficulty one faces theoretically when an attempt is made to model cooperative phenomena in lipid membranes is that modelling of large assemblies of molecules with complex internal conformations and of molecular systems with translational degrees of freedom in two dimensions are each very complex problems to treat individually. A combined treatment seems at present prohibitive. Hence, three routes have been taken by theorists. One is to include both translational and conformational variables in as much detail as possible and then perform a simulation study on only a small number of molecules, excluding a reliable treatment of the cooperative phenomena [8–15]. Another route is to take accurate account of the conformational variables and deal with the translational variables in a very approximate manner, for example, via a lattice-gas Potts model which only models the polycrystalline and granular structure of the solid phase and neglects the subtleties of two-dimensional melting [16–19]. A third route mediates these two by dealing with the internal state variable in a very approximate way [20–23]. This approximation permits a reasonably large number of molecules to be treated in a simulation study and hence leads to information about the transition properties.

In the present paper we shall describe results obtained from models which proceed along the second and third routes. These models include a two-dimensional lattice-gas Potts model (the Pink-Potts model) and a hard-disc model of discs with variable sizes, cf. Fig. 2. The lattice-gas Potts model is studied by conventional Metropolis Monte Carlo lattice simulation techniques [6] well-known from spin systems. The hard-disc model is studied by constant $NpT$ Monte Carlo simulations.

## 2. The lattice gas Pink-Potts model: a model with true internal and approximate translational degrees of freedom

The most detailed information presently available from theoretical studies of lipid bilayers and monolayers has been obtained from statistical mechanical lattice models [4–6]. Among the most successful lattice models is Pink's multi-state model of the chain-melting phase transition. Building upon some early ideas of Doniach [24] and Marčelja [25], the Pink model takes detailed account of the acyl-chain conformers, their rotational isomeric statistics, as well as their mutual interactions. The lattice approximation, which automatically takes care of the excluded volume interactions, ignores the translational degrees of freedom. The Pink model considers the lipid bilayer as two non-interacting independent monolayers, each of which is represented by a two-dimensional triangular lattice, as shown in Fig. 2(a). The two chains of the diacyl lipid molecule are furthermore assumed to be independent and the carbonyl groups of the chains are restricted to lie in the plane of the lattice. The basic idea underlying the Pink model is to map the three-dimensional acyl-chain conformational variables upon a finite, discrete set of projected two-dimensional coarse-grained variables, $(A_n, E_n, D_n)$, $n = 1, 2, ..., q$, where $n$ is a label of the $q$ states. Each state is characterized by a cross-sectional area, $A_n$, an internal conformational energy, $E_n$, and a single-chain degeneracy, $D_n$. The degeneracy reflects the fact that the mapping from the three-dimensional variables to the two-dimensional variables is a many-to-one relation. The number of states, $q$, included in the multi-state model depends on the level of detail required. Most of the model studies reported so far use ten states, $q = 10$. Recently, the full single-chain density of states has been worked out and used within a Pink-type model [26]. The states entering the ten-state model are chosen [5] by imposing conditions of low conformational energy and optimal packing. Drawing upon concepts developed for inter-molecular interactions in anisotropic liquids, the interaction energy between the hydrophobic parts of nearest-neighbor acyl chains within the Pink model is assumed to take the form

$$E_{nm} = -J_o V_{nm} S_n S_m, \qquad (1)$$

where $J_o$ is a van der Waals interaction constant ($J_o \sim 10^{-13}$ erg), $V_{nm}$ describes the distance dependence of the potential, and $S_n$ is the nematic shape factor or so-called average acyl-chain order parameter (segmental order parameter)

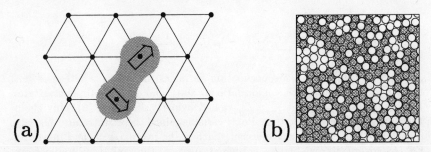

Fig. 2. (a) Schematic illustration of the lattice-gas Pink-Potts model. Sites of the triangular lattice are occupied by lipid molecules with a schematic indication of the areas of the two acyl chains. The Potts variable is denoted by an arrow. (b) Schematic illustration of a binary hard-disc model.

$$S_n = \frac{1}{2(N-1)} \sum_{i=2}^{N} (3\cos^2 \vartheta_{ni} - 1). \qquad (2)$$

The summation extends over all segments of the chain and $\vartheta_{ni}$ is the angle between the bilayer normal and the normal to the plane spanned by the $i$th CH$_2$-group of the $n$th conformational state of the chain. For the completely ordered all-*trans* state (the ground state), $S_1 = 1$. The dependence of the interaction potential is taken to be

$$V_{nm} \sim r_{nm}^{-5} \sim (A_n A_m)^{5/4}. \qquad (3)$$

Of the ten single-chain states, the nine lower ones (including the ground state) are characteristic of the low-temperature phase and the tenth state is a conformationally very disordered, fluid state. It becomes obvious at this point that the Pink model is not a lattice model in a strict sense, but rather a topological structure which assures that each chain interacts with six neighbors corresponding to a close-packed arrangement. The chains are simply assigned different amounts of space depending on their conformational state. The description of the Pink model is completed by adding a term representing the pair interaction between the polar head groups. This may be done by a simple Coulomb-type interaction which, however, can effectively be accounted for, as first suggested by Marčelja [25], by an intrinsic lateral pressure, $\Pi$, which couples to the total membrane area.

An extension of the Pink model has been proposed [18] to model, in the simplest possible setting, the effects of crystallization processes on the chain-melting transition. On a lattice this is done artificially by introducing a set of variables, formally Potts variables, which for each site label the orientation of the crystalline domain in which the chain at that site prefers to participate. Molecules in the same microcrystalline domain assume the same value of their Potts variable and have an attractive interaction. Molecules in differently oriented domains repel each other corresponding to an amount of energy $J_P$. Such an interaction models crystalline misfit and packing defects and is related to a grain-boundary potential pertinent for a polycrystalline material [27]. The Potts interaction is hence assumed to be isotropic (step-function) and the number, $Q$, of Potts states is taken to be very large (e.g., $Q = 30$). The full Hamiltonian for the combined Pink-Potts model may now be written

$$\mathcal{H} = \mathcal{H}_{\text{Pink}} + \mathcal{H}_{\text{Potts}}, \qquad (4)$$

where

$$\mathcal{H}_{\text{Pink}} = \sum_i \sum_{n=1}^{10} E_n \mathcal{L}_{in} - \frac{J_o}{2} \sum_{\langle i,j \rangle} \sum_{m,n} V_{mn} S_m S_n \mathcal{L}_{mi} \mathcal{L}_{nj} + \Pi \sum_i \sum_{n=1}^{10} A_n \mathcal{L}_{ni} \qquad (5)$$

and

$$\mathcal{H}_{\text{Potts}} = \frac{J_P}{2} \sum_{\langle i,j \rangle} \sum_{m,n} \sum_{p,r=1}^{9} (1 - \delta_{pr}) \mathcal{L}_{ip}^P \mathcal{L}_{jr}^P \mathcal{L}_{im} \mathcal{L}_{jn}. \qquad (6)$$

In Eq. (6) $\mathcal{L}_{in} = 0, 1$ and $\mathcal{L}_{ip}^P = 0, 1$ are occupation variables indicating whether or not the chain at the $i$th site is in the $n$th conformational state and in the $p$th Potts state, respectively, that is, $\sum_{n=1}^{10} \mathcal{L}_{in} = 1$ and $\sum_{p=1}^{Q} \mathcal{L}_{ip}^P = 1$. Obviously, from Eq. (6), only chain states among the nine gel states couple their Potts variables since these states are the only ones which can pack into a crystalline environment. Each of the two parts of the Pink-Potts Hamiltonian, Eq. (4), governs its own first-order phase transition: the Pink Hamiltonian from a conformationally ordered (o) state to a conformationally disordered (d) state and the Potts Hamiltonian from a Potts ordered (crystalline solid, s) state to a Potts disordered (liquid, l) state. Depending on the coupling ratio, $J_P/J_o$, these two transitions may be coupled or not as shown on the mean-field phase diagrams in Fig. 3. For coupling ratios below the

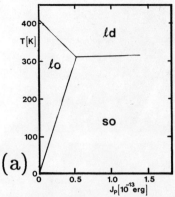
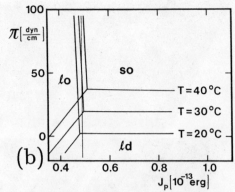

**Fig. 3.** Mean-field phase diagrams of the Pink-Potts model. (a) is pertinent for thermally driven transitions in a lipid bilayer and (b) for pressure-driven transitions in a lipid monolayer. The phase labels are denoted **lo** (liquid-ordered), **so** (solid-ordered), and **ld** (liquid-disordered).

triple point, the two transitions are decoupled leading the way to an intermediate phase, a liquid-ordered (**lo**) phase, in between the solid-ordered (**so**) and the liquid-disordered (**ld**) phase.

Before we describe simulation results for the full Pink-Potts model we first refer to some results for the plain Pink model in order to illustrate the fact that the chain order-disorder transition, despite its inherent first-order character, is strongly influenced by lateral thermal density fluctuations [6,28]. These density fluctuations manifest themselves in dramatic peaks in the response functions, such as the specific heat and the lateral compressibility, and in almost continuous variations of properties like the order parameter (membrane area) and the internal energy [29,30]. This is in close agreement with a large body of experimental data (for references to the experimental work, see [6]). These observations may be described in terms of pseudo-critical behavior and apparent power-law singularities [30,31]. The pseudo-spinodal points are very close ($\sim 0.001 T_m$) to the equilbrium transition point. Hence the chain-melting transitions in lipid bilayers are very close to a critical point, $T_m$. Microscopically, the density fluctuations lead to cluster formation phenomena. We shall focus here on these phenomena since they are physically interesting as well as relevant for membrane function. In Fig. 4 are shown snapshots of microconfigurations as obtained from Monte Carlo simulations of the Pink model for three different lipid membranes [31] near their chain-melting transition temperature. These three lipids, DMPC, DPPC, and DSPC, are very similar and only differ with respect to acyl-chain length (14, 16, and 18 carbons, respectively). It is clear that the transition in all three systems is associated with strong fluctuations and that the fluctuations are enhanced as the chain length is decreased and the system is driven close to the critical point. We have calculated as a function of temperature the full cluster-size distribution function, $n_\ell^\alpha(T)$, where $\alpha$ is a phase label and $\ell$ is the number of chains in the cluster. Since the transition is of first order, $n_\ell^\alpha(T)$ has obviously a cut-off at some finite value of $\ell$. Nevertheless, due to pseudo-criticality, the cluster-size distribution function has some interesting scaling properties [32] below this cut-off as illustrated in Fig. 5(a) in the case of DPPC bilayers. This figure gives a double logarithmic plot of $n_\ell(T)$ for a temperature 0.2% above the equilibrium transition temperature. This figure shows that for cluster sizes up to the largest ones, an effective scaling relation

$$n_\ell \sim \ell^{-\tau} \qquad (7)$$

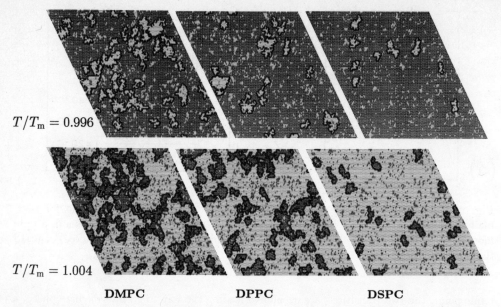

$T/T_m = 0.996$

$T/T_m = 1.004$

DMPC  DPPC  DSPC

**Fig. 4.** Snapshots of microscopic configurations of the Pink model for the lipid bilayer chain-melting transition as obtained from Monte Carlo calculations for three different lipid species near their respective transition temperature $T_m$. The bulk phases, the clusters, and the interfaces are shown schematically. The system contains $100 \times 100$ chains on a triangular lattice.

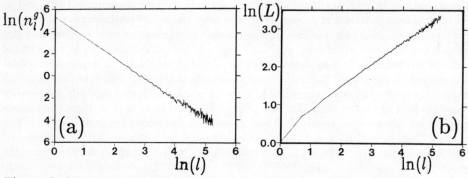

**Fig. 5.** Scaling properties of the clusters observed in Monte Carlo simulations of the Pink model for DPPC lipid bilayers, from Eqs. (7) and (8).

holds with $\tau \simeq 1.84$. The same relation holds immediately below $T_m$ as well as for the other lipid systems. For temperatures further from $T_m$, the scaling relation holds only for clusters up to a certain size from where a crossover is found. Eq. (7) is only a pseudo-critical scaling relation. A similar relationship with almost the same value of the exponent was recently found to apply to the two-dimensional five-state ferromagnetic Potts model at its first-order transition point [33]. The $Q$-state Potts model with $Q$ just above the critical dimension ($Q_c = 4$ in two dimensions) is known to display pseudo-critical behavior. At a real critical point, for example, as in the two-dimensional Ising model, the scaling relation Eq. (7) is known to hold with $\tau \simeq 2.1$ being the exponent of two-dimensional percolation [34]. Another scaling property derives from the morphology of the clusters. By assigning to each cluster a linear dimension, for example, from the radius of gyration, and by using the simulated cluster statistics to determine an average cluster diameter, $L$, we obtain the data in Fig. 5(b). This plot reveals a scaling relation of fractal geometry

$$\ell \sim L^D \tag{8}$$

with $D \simeq 1.67$. For clusters at a critical point, $D$ is the fractal exponent of percolation clusters, that is, $D \simeq 1.9$ [34]. To conclude our brief discussion of fluctuation phenomena and pseudo-criticality of lipid membranes we note that the exponents derived from analyses of the Monte Carlo response functions in terms of power-law singularities assume their classical (mean-field) values and these values apply to different lipid systems. This is in accordance with a variety of experimental measurements (see, for example, [31]). These results, together with those reported above for the scaling properties of the clusters, support the idea of universality in the pseudo-critical behavior of lipid membrane phase transitions.

From the point of view of membrane science the finding from computer-simulation work of dynamically heterogeneous membrane states, (Fig. 4), induced by pseudo-critical lateral density fluctuations is very important since it leads to a proposal for a simple and general physical mechanism underlying diverse phenomena as transmembrane permeation, exchange of molecules between different membranes, and the activity of certain interfacially active enzymes [6,31,32,35]. The basic idea behind this general mechanism is that the cluster formation implies a concomitant formation of a fluctuating interfacial environment bounding the clusters and the bulk matrix, as can be seen in Fig. 4. These interfaces have special packing properties. As an example of the predictive power of the simulation of the Pink model we show in Fig. 6 the results for the transmembrane permeability of $Na^+$ ions as compared to appropriate experimental data [36,37]. The theoretical curve is derived on the basis of a very simple model which assigns the full temperature dependence of the permeation to that of the interfacial measure. The agreement is very good.

We now turn to report on simulation results for the full Pink-Potts model [16–18], first in the case of the bilayer phase behavior. For the bilayer, the two transitions are expected from experiments to be coupled in equilibrium [38] although they may not be so out of equilibrium since the acyl-chain ordering process is much faster than the crystallization process. The non-equilibrium properties of the Pink-Potts model have been studied by Monte Carlo simulations (in the case of coupled transitions) by performing thermal cycling

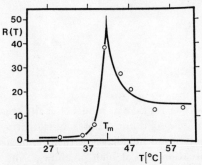

**Fig. 6.** Relative transmembrane permeability of $Na^+$ ions through DPPC lipid bilayers as measured experimentally (◯) [37] and as calculated in Monte Carlo simulations of the Pink model.

through the transition region [16,18]. The interesting findings from these simulations include the phenomenon of interfacial melting [16] which is a non-equilibrium process by which the Potts domain boundaries in a Potts-ordered chain-ordered (**so**) polycrystalline aggregate melt as the temperature is raised towards the bulk melting point subsequent to a thermal quench. This process is related to the phenomenon of grain-boundary melting in alloys as well as interfacial adsorption, surface melting, and wetting [39]. Under appropriate conditions, a non-equilibrium **lo** phase may form subsequent to rapid cooling. This phase is a gel phase with conformationally ordered chains but otherwise structurally disordered (amorphous or Potts disordered).

An even more interesting scenario can arise in lipid monolayers, which in some cases are known to display an intermediate phase [17,19,40]. In Figs. 7 and 8 are shown the Monte Carlo data for a DPPC monolayer isotherm and some typical microconfigurations obtained along this isotherm. The data were obtained from the Pink-Potts model in the case of decoupled transitions (Fig. 3(b)). The main branch of the isotherm (solid line) is obtained from a series of increasing lateral pressures starting from the **ld** phase. Each point is equilibrated for 4000 MCS/S. For this equilibrium time the isotherm is almost reversible in the **lo** and **ld** phases with only a slight hysteresis around the **ld-lo** transition. The relaxation in the transition region is very slow, causing the isotherm to be nonhorizontal [28]. In the **so** phase, however, the ordering kinetics is extremely slow and the equilibrium state cannot be attained for the $100 \times 100$ system within the maximum observation time ($t \leq 8000$ MCS/S). For a smaller system, $N = 30 \times 30$, it is possible to make the system go into the stable uniformly ordered Potts phase at high pressures. This is illustrated by the dotted line in Fig. 6. The calculations show that the **lo-so** transition is associated with a discontinuity in molecular area of approximately 0.8 Å$^2$. The remaining branch on Fig. 6 (the dashed line) corresponds to an isothermal scan performed by starting from a high-pressure uniformly ordered Potts phase. It turns out to be impossible to make the system enter the Potts disordered phase **lo** for any positive lateral pressure, even for the smallest system studied, $N = 30 \times 30$. Hence, the entire low-$\Pi$ part of this branch corresponds to a metastable

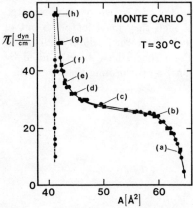

**Fig. 7.** Monte Carlo isotherm obtained for a lipid monolayer Pink-Potts model. Results are shown for systems with 10000 (□) and 900 (○) chains. The solid line denotes an isotherm which at high pressures ($\Pi$) corresponds to a non-equilibrium monolayer with a domain structure. The dotted line refers to an equilibrium situation also at high pressures. The dashed line denotes at low pressures a metastable monolayer with a single uniformly Potts-ordered (crystalline) domain. Points (a)–(h) on the isotherm refer to the microconfigurations in Fig. 8.

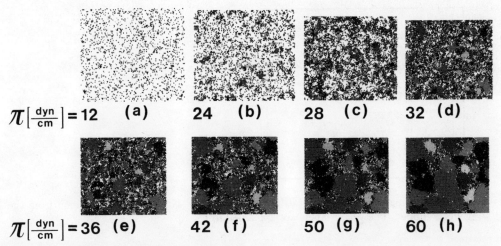

**Fig. 8.** Snapshots of microconfigurations for the points (a)–(h) on the isotherm of Fig. 7. The configurations refer to Monte Carlo simulations of lattices with $100 \times 100$ chains. White areas indicate fluid domains and symbols indicate acyl chains in gel-like conformations, with each symbol labelling a Potts state.

situation. The lateral pressure is simply a thermodynamic variable which is too weak to appreciably affect the ordering dynamics within a reasonable time. From these observations we conclude that the high-$\Pi$ shoulder of the monolayer isotherm is strongly dependent on non-equilibrium effects. The longer the system is allowed to equilibrate in this region of the phase diagram, the less rounded the high-$\Pi$ shoulder will be. The time evolution of this rounding may be interpreted in terms of a grain-growth process.

The microscopic phenomena underlying the various ordering processes which occur as the monolayer is taken along the isotherm are illustrated in Fig. 7. The phenomena which are examples of microscopic pattern formation, non-equilibrium crystallization, and grain growth illustrate the intricate interplay between the two sets of variables of the Pink-Potts model. Snapshots of direct representations of microconfigurations characteristic of selected lateral pressures are shown in this figure. For pressures below the **lo-so** transition, these snapshots at higher pressures characterize non-equilibrium configurations for a system with a lateral-pressure history corresponding to 4000 MCS/S at each successive pressure of an increasing series as given in Fig. 6. For increasing $\Pi$, the microconfigurations reflect the dramatic compression at the **ld-lo** transition where the acyl chains undergo the conformational ordering transition into the Potts disordered solid state **lo**. Upon further increase in pressure the system contracts only very slightly. The approach to the **so** phase is signalled by the appearance of clusters of chains in the same Potts state. These clusters are akin to density fluctuations. The cluster dynamics are extremely slow as the pressure changes. When the **lo-so** phase line is crossed, these clusters grow persistently and eventually at long times or high pressures form the pattern of a polycrystalline aggregate characterized by a random network of grain boundaries separating competing domains with different Potts states. Due to the high degeneracy of the 30-state Potts model, the grains are compact. The domain pattern anneals extremely slowly in time. This is due to the fact that the lateral pressure only couples to the Potts ordering via the fluid chains which have almost disappeared in the highly condensed **lo** and **so** phases. There are even domains of the **lo** phase which are intercalated between Potts ordered grains. Moreover, the boundaries between the various ordered grains are typically widened (softened) by an interfacial region of Potts disorder.

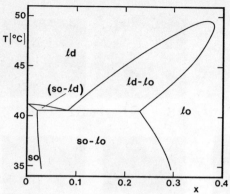

**Fig. 9.** Phase diagram of the lipid (DPPC)-cholesterol bilayer system as obtained from the Pink-Potts model. The phase labels are explained in Fig. 3.

This disordered interfacial network screens the interaction between the ordered domains, lowers the driving force for grain growth, and consequently slows down the growth kinetics.

So far we have restricted ourselves to the modelling of pure one-component systems. New phenomena arise as we consider the effects on the phase behaviour of a second component, an impurity. Relevant 'impurities' in membranes may be cholesterol, proteins, or anesthetics [32]. The case of cholesterol is a particularly interesting one since its treatment lies right at the heart of the coupling between the two lipid molecule degrees of freedom [19,35,41]. Cholesterol is an amphiphilic molecule but in contrast to the lipids its hydrophobic moiety consists of a stiff sterol skeleton with very little flexibility. Hence it can be considered a hydrophobically smooth object with no internal degrees of freedom. The peculiar phase diagram of lipid-cholesterol mixtures, (Fig. 9), may be rationalized by the Pink-Potts model in which the lipid-cholesterol interactions are included [41]. These interactions reflect the fact that the cholesterol molecule has mixed feelings about the lipid phases: on the one hand the cholesterol molecule prefers to have conformationally ordered lipid chains next to it, but on the other hand the cholesterol molecule does not fit very well into the solid (crystalline) lipid bilayer phase in which the ordered chains prevail. Hence, cholesterol is frustrated—it couples differently to the two diferent lipid degrees of freedom. The way cholesterol copes with this is to break the crystal and fluidize the lipid gel phase, without lowering the acyl-chain conformational order. In the language of the Pink-Potts model, cholesterol decouples the crystalline (Potts) and the conformational degrees of freedom. Turning to the lateral density fluctuations, computer simulations of the Pink model [35] have shown that the mobile dispersion of cholesterol (annealed impurities) in the membrane for low cholesterol concentrations lead to an enhancement of the fluctuations and an increase of the average cluster size and the interfacial measure [35]. This is in accordance with a number of experiments (see, for example, [42]) and has some dramatic consequences for the capacity of cholesterol to act as a membrane sealer. Furthermore, it shows that cholesterol is a potent agent for modulating interfacial fluctuations.

Another interesting membrane 'impurity' is the large class of general anesthetics, which are typically small molecules that are soluble in the membrane as well as in the surrounding aqueous phase. The physical effects of anesthetics on membranes may be modelled by the Pink model extended to incorporate a lattice of interstitial sites which can be occupied by the anesthetic molecules [32]. Since the small molecules can exchange to the water phase, the model has to be simulated within the grand canonical ensemble in which the anesthetics concentration is governed by a chemical potential in the water phase. The simulations show that these circumstances have a dramatic effect on the phase equilibria and the density fluctuations [32]. For example it has been found that certain anesthetics suppress the density fluctuations and accumulate in the interfacial environment. These findings may have some bearing on the action mechanisms of general anesthetics.

## 3. The hard disc mixture: a model with true translational and approximate internal degrees of freedom

Recently a new technique has been developed to study a hard-disc model of the lipid membrane phase transition [23], in particular the pressure-induced monolayer transition. The model includes true translational degrees of freedom and approximates the internal variables by allowing for different disc sizes as well as conversions between different sizes. The model may be considered a minimal model of lipid membranes in the sense that it only includes the most basic physics and neglects details of the conformational states and their detailed statistics as well as the soft interaction between the molecules. In this respect it differs from the model of Scott and Cheng [43]. In the simplest version of the model only two disc sizes, $\sigma_1$ and $\sigma_2$ ($\sigma_1 > \sigma_2$) are allowed. The two disc sizes are assigned different degeneracies, $D_1$ and $D_2$, with $D_1 > D_2$. The small discs resemble molecules with an internal state corresponding to conformational order and the large discs resemble molecules with an internal state corresponding to conformational disorder. The composition of this binary disc mixture is not an independent variable but determined by the thermodynamic conditions. Hence we do not have fixed composition. The binary mixture is described by two independent variables, $\chi = \sigma_1/\sigma_2$ and $\nu = D_1/D_2$. For this mixture the following scenario of phase transformations is possible: (i) an order-disorder transition among discs of the same size (the solid-liquid transition well-known to occur in hard-core systems [44] (note that there is still some controversy regarding the nature of this transition [45])), and (ii) a transition from large discs to small discs with increasing pressure (the 'large-small' transition).

The computer simulations of the model follow the Metropolis Monte Carlo sampling procedure for the $NpT$ ($N\Pi T$) ensemble. The random particle moves obey the hard-core conditions. Binary mixtures of discs require a further move consisting of a change in disc size. For a given disc of size $\sigma_i$, a trial disc size, $\sigma_j$, is chosen with probability $P(j)$ where

$$P(j) = \frac{D_j}{D_1 + D_2}. \tag{9}$$

Moves which would result in overlap are rejected. A change of disc size is attempted before each particle move. As part of the constant $NpT$ simulations it is necessary to uniformly scale the system. A random change in the simulation box size, $\delta L$, is considered where $\delta L$ is given by

$$\delta L = (2\zeta - 1)\delta L_{\max}, \tag{10}$$

with random number $\zeta$, $0 \leq \zeta \leq 1$, and maximum change in box length $\delta L_{\max}$. $\delta L_{\max}$ is independently adjusted to give an acceptance ratio of 50% for the area moves. The probability of accepting a random change from state $i$, at area $A(i) = L_i^2$, to state $j$, at area $A(j) = L_j^2 = (L_i + \delta L)^2$, is given by $\min(1, P_{ij})$ where

$$P_{ij} = \exp(-\delta H_{ij}/k_B T), \tag{11}$$

$$\delta H_{ij} = \Pi \Delta A - Nk_B T \ln[A(j)/A(i)]. \tag{12}$$

Here $\Pi$ is the two-dimensional or lateral pressure. Moves which would result in overlap are rejected. A scaling of the system is attempted once every complete Monte Carlo cycle (one attempted move per particle).

As an example of the kind of results which may be obtained by this technique we focus on a system with $\chi = 1.15$ and $\nu = 99$. A system with 400 discs is considered with equilibration runs of 5000 cycles and production runs of 10000 cycles. Snapshots of some of the configurations, at different reduced pressures, are shown in Fig. 10. There are relatively few small discs in the system right up to the start of the liquid-solid transition and all the systems are disordered. Within the liquid-solid transition region (Figs. 10(d) and (e)) the large discs form more-ordered domains. The small discs are found between these domains and within less-ordered, fluid-like regions. At densities slightly higher than that corresponding to the packing fraction, $\eta_{sol} = 0.73$, of the solidus, Fig. 10(f), the ordered domains are much larger and many of the small discs take on the same order as the large discs. There are still a few disordered regions within the systems. The number of small discs starts to increase noticeably once the structure becomes almost fully ordered. At the highest densities there are few large discs in the systems. The density-concentration profile is shown in Fig. 11(a). The inset shows the same data on a semi-log plot. Within the transition region the number of large discs actually increases slightly. Then, as the density is increased, the concentration drops rapidly. The start of this drop corresponds to the density at which the systems are fully ordered, though they do contain defects. The sharpness of the drop is most apparent on the semi-log plot. In response to the applied pressure, the system wants to reduce its area. It can do this either by reducing its "free area", which requires an increase in its order at these densities, or by reducing the total area of the discs and leaving the free area the same. Across the liquid-solid transition region the increase in order within the large discs provides sufficient free area also to allow an increase in the number of large discs. The number then reduces gradually while the amount of order in the system increases further. Once the system becomes fully ordered, however, the only way to reduce the area is to reduce the number of large discs. The $\Pi$-$A$ isotherm is shown in Fig. 11(b). The lower dashed line in the figure is the so-called base curve, which describes the behavior of a system with a single disc size. The upper dashed curve is the same curve but scaled in both axes so that it corresponds to a system with disc diameters of $1.15\sigma$. The Monte Carlo data can be divided into three

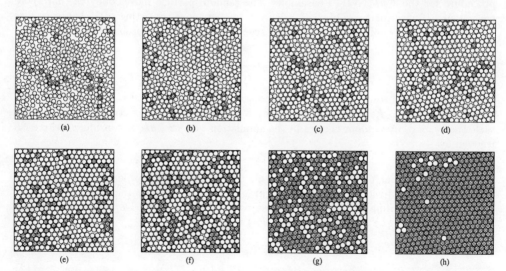

**Fig. 10.** Snapshots of configurations of mixtures of 440 hard discs, with $\nu = 99$ and $\chi = 1.15$, at pressures, $p/k_B T$, of (a) $2.79\sigma^{-2}$, (b) $5.86\sigma^{-2}$, (c) $7.14\ \sigma^{-2}$, (d) $8.51\ \sigma^{-2}$, (e) $10.5\sigma^{-2}$, (f) $12.0\sigma^{-2}$, (g) $13.9\sigma^{-2}$, and (h) $16.4\sigma^{-2}$. The systems are scaled to the same box size and the shaded regions indicate the Voronoi polygons associated with the smaller discs.

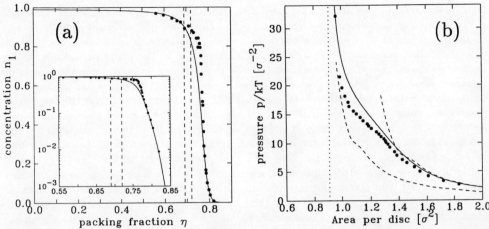

**Fig. 11.** Monte Carlo results for a binary hard disc mixture of 400 discs with $\nu = 99$ and $\chi = 1.15$. (a): The concentration, $n_1$, of large discs, as a function of packing fraction $\eta$. The dashed lines give the liquid-solid transition region for a one-component system. The solid line is scaled particle theory [23]. (b): Lateral pressure as a function of area per disc. The points (•) are Monte Carlo data, the solid line is data from scaled particle theory. The dashed lines are the base curve and the same curve scaled to correspond to a system with disc diameters of $1.15\sigma$. The dotted line is the high density limit, $A = \pi/2\sqrt{3}$.

regions. At low densities the data follow the scaled base curve, but at a slightly reduced pressure. At high densities the data join and then follow the base curve. In the intermediate region the data follow an almost straight path from the scaled base curve to the true base curve. This is the transition region for a transition from a predominantly large-disc system to a predominantly small-disc system. The low-density limit of the large-small transition region, as apparent from the $\Pi$-$A$ isotherm, corresponds to the density at which the system first becomes fully ordered. It should be noted that there is no evidence in this plot to suggest the presence of a liquid-solid, order-disorder transition.

The parameter $\chi$ determines the density at which the concentration of large discs within the systems first tends to zero. The value of $\nu$ determines the rate of the transition from large discs to small discs. If this transition occurs at densities lower than $\eta_{\text{liq}} = 0.69$, then it is completely decoupled from the liquid-solid transition. The small-disc systems will subsequently behave as if there were only one species of disc. If, on the other hand, the large-small transition spans a range of densities including the liquid-solid transition, then their relative positions may be important. It was shown above that the increase in order within the domains of large discs at densities just into the liquid-solid transition region also caused an increase in the concentration of large discs. This tends to accentuate the "plateau" in the isotherm. Systems with a lower degeneracy ($\nu = 19$) but with the same $\chi$ value, for which the two transitions are also coincident, show no noticeable increase. The large-small transition for systems with a very high degeneracy (for example, $\nu = 10^5$) coupled with a relatively large difference in disc sizes ($\chi = 2$) spans a range of densities almost coincident with that of the liquid-solid transition. In this case the "plateau" in the isotherm is quite pronounced and interesting hysteresis effects occur as a result of the high degeneracy.

The hard-disc models studied here open up a novel and promising way of modelling lipid monolayer phase behavior. Here we only outline the perspectives of such an approach by proposing the binary mixture model with $\chi \simeq 1$–$2$ and $\nu \gg 1$ as a minimal model of a lipid monolayer. In this model, the large disc size corresponds to the extended (melted, conformationally disordered) acyl-chain state and the small disc size corresponds to the all-*trans* (conformationally ordered) acyl-chain state. The large value of $\nu$ reflects the fact that the melted state carries a large internal entropy due to the many excitations of the disordered chain. The model is minimal in the sense that it only explores the effects of excluded volume and internal degrees of freedom and neglects direct, softer interactions. Comparison of the model isotherms (Fig. 11(b)) with experimental isotherms [46] for DPPC monolayers, suggests that the minimal model has captured the essential physics and hence underscores an earlier assertion by Nagle [4] that the excluded volume interactions are the dominant factor for the phase behavior. Moreover, it is interesting to note that the minimal model confirms some of the results of the Pink-Potts lattice-model calculations (Fig. 7) that the formation of a granular non-equilibrium structure with well-defined grain boundaries (Figs. 8 and 10) is important for interpreting the shape of the high-pressure experimental isotherms. In the minimal model, due to the lack of direct, soft interactions, the solid-fluid transition and the large-small transition are not coupled, although the parameters $\chi$ and $\nu$ may be tuned so that the transitions almost coalesce. Hence one has to await studies of extensions of the model which include soft interactions in order to relate the theory to the experimental finding of an intermediate phase in some lipid monolayer systems [40]. This intermediate phase is characterized by chain conformational order and positional disorder (that is, a small disc liquid in the language of the minimal model).

## 4. Conclusions

In this paper we have given a brief review of recent activity in the field of theoretical modelling of phase transitions in two-dimensional systems of molecules with internal degrees of freedom using simulation techniques. The particular systems considered are lipid membranes, bilayers and monolayers, which are a unique class of physical systems which possess two different mechanisms for ordering and condensation, one being the conformational ordering of the acyl chains via rotational isomerism (the internal degree of freedom), and the other being crystallization involving translational degrees of freedom. These two sets of degrees of freedom and the couplings between them impart to the membrane a rich phase behavior, including the possibility of condensation in one or both sets of variables, separately or simultaneously. These complications are reflected in the considerable difficulties encountered in constructing realistic theoretical models of lipid monolayer phase behavior, the basic problem being that, if both sets of degrees of freedom and their interactions are modelled as realistically as possible, it is difficult to make calculations in the transition region. Therefore, the most successful theoretical studies of membrane phase behavior to date are based on lattice models which represent the acyl-chain conformational variables accurately, but treat the lateral degrees of freedom approximately via coarse-grained crystalline variables. The Pink-Potts lattice gas model discussed in Sec. 2 is such a model.

A more recent and realistic attempt to model lipid bilayers with approximate conformational and true translational degrees of freedom is that discussed in Sec. 3. Here a minimal model of lipid-membrane phase behavior is considered which takes hard-core repulsion between lipid molecules as the basic molecular interaction. The model is formulated in terms of two-dimensional variables and the lipid molecules are assumed to be hard discs of various sizes reflecting the different internal conformational states of the chains. It turns out that this model, which faithfully accounts for the packing properties of a fluid system, reproduces all the fundamental features of the chain-melting transition and hence shows that the excluded volume interaction is the dominant factor for the lipid bilayer chain-melting transition. It is reassuring to note that the non-equilibrium states found in the hard-disc model resemble the heterogeneous membrane states of the Pink-Potts model. The class of models for mixtures of hard-discs studied in this paper is expected to have a general sphere of applicability. So far we have only explored the properties of the models in the case of a binary mixture, but extensions to multi-component mixtures are straightforward, although not necessarily easy to study by Monte Carlo simulation due to the poly-dispersivity.

## Acknowledgements

This work was supported by NSERC of Canada, by FCAR du Quebec, by the Danish Natural Science Research Council under grants J.nos. 5.21.99.71, 11-7498, and 11-7785, and by Løvens Kemiske Fabriks Fond. Stimulating discussions with Søren Toxværd and Eigil Præstgaard are gratefully acknowledged.

## References

1. O.G. Mouritsen, in *Physics in Living Matter*, (D. Baeriswyl, M. Droz, A. Malapinas, and P. Martinoli, eds.) Springer-Verlag, Heidelberg, 1987, p.76.
2. C. Cevc and D. Marsh, *Phospholipid Bilayers. Physical Principles and Models*. John Wiley and Sons, New York, 1987.
3. B. Huberman, D. Lublin, and S. Doniach, Solid St. Commun. **17**, 485 (1975).
4. J.F. Nagle, Ann. Rev. Phys. Chem. **31**, 157 (1980).
5. A. Caillé, D. Pink, F. de Verteuil, and M.J. Zuckermann, Can. J. Phys. **58**, 581 (1980).
6. For a recent review on microscopic modelling and computer simulation of lipid membrane cooperative phenomena, see O.G. Mouritsen, in *Molecular Description of Biological Membrane Components by Computer Aided Conformational Analysis* (R. Brasseur, ed.) CRC Press, Boca Raton, Florida, 1990, p.3.
7. M. Bloom and O.G. Mouritsen, Can. J. Chem. **66**, 706 (1988).
8. P. van der Ploeg and H.J.C. Berendsen, Mol. Phys. **49**, 233 (1983).
9. H.L. Scott, J. Chem. Phys. **80**, 2197 (1984).
10. H.L. Scott, Biochemistry **25**, 6122 (1986).
11. E. Egberts and H.J.C. Berendsen, J. Chem. Phys. **89**, 3718 (1988).
12. H.L. Scott, Phys. Rev. A **37**, 263 (1988).
13. S.H. Northrup and M.S. Curvin, J. Phys. Chem. **89**, 4707 (1985).
14. J.P. Bareman, G. Cardini, and M.K. Klein, Phys. Rev. Lett. **60**, 2152 (1988).
15. J. Harris and S.A. Rice, J. Chem. Phys. **89**, 5898 (1989).
16. O.G. Mouritsen and M.J. Zuckermann, Phys. Rev. Lett. **58**, 389 (1987).
17. O.G. Mouritsen and M.J. Zuckermann, Chem. Phys. Lett. **135**, 294 (1987).
18. M.J. Zuckermann and O.G. Mouritsen, Eur. Biophys. J. **15**, 77 (1987).
19. J.H. Ipsen, O.G. Mouritsen, and M.J. Zuckermann, J. Chem. Phys. **91**, 1855 (1989).
20. Z.-Y. Chen, J. Talbot, W.M. Gelbart, and A. Ben-Shaul, Phys. Rev. Lett. **61**, 1376 (1988).
21. R.M.J. Cotterill, Biochim. Biophys. Acta **433**, 264 (1976).
22. D.P. Fraser, R.W. Chantrell, D. Melville, and D.J. Tildesley, Liquid Cryst. **3**, 423 (1988).
23. D.P. Fraser, M.J. Zuckermann, and O.G. Mouritsen, Phys. Rev. A **42** (in press, 1990).
24. S. Doniach, J. Chem. Phys. **68**, 4912 (1978).
25. S. Marčelja, Biochim. Biophys. Acta **367**, 165 (1974).
26. F.P. Jones, P. Tevlin, and L.E.H. Trainor, J. Chem. Phys. **91**, 1918 (1989).

27. G.S. Grest, M.P. Anderson, and D.J. Srolovitz, Phys. Rev. B **38**, 4752 (1988).
28. O.G. Mouritsen, J.H. Ipsen, and M.J. Zuckermann, J. Coll. Int. Sci. **129**, 32 (1989).
29. O.G. Mouritsen, Biochim. Biophys. Acta **731**, 217 (1983).
30. O.G. Mouritsen and M.J. Zuckermann, Eur. Biophys. J. **12**, 75 (1985).
31. J.H. Ipsen, K. Jørgensen, and O.G. Mouritsen, Biophys J. (in press, 1990).
32. O.G. Mouritsen, K. Jørgensen, J.H. Ipsen, M.J. Zuckermann, and L. Cruzeiro-Hansson, Phys. Scr. ( in press, 1990).
33. P. Peczak and D.P. Landau, Phys. Rev. B **39**, 11932 (1989).
34. D. Stauffer, Phys. Rep. **54**, 1 (1979).
35. L. Cruzeiro-Hansson, J.H. Ipsen, and O.G. Mouritsen, Biochim. Biophys. Acta **979**, 166 (1989).
36. L. Cruzeiro-Hansson and O.G. Mouritsen, Biochim. Biophys. Acta **974**, 63 (1988).
37. D. Papahadjopoulos, K. Jacobsen, S. Nir, and T. Isac, Biochim. Biophys. Acta **311**, 330 (1973).
38. M. Caffrey, Biochemistry **24**, 4826 (1985).
39. O.G. Mouritsen, H.C. Fogedby, E.S. Sørensen, and M.J. Zuckermann, in *Time-Dependent Effects in Disordered Materials* (R. Plynn and T. Riste, eds.) Plenum Publ. Co., New York, 1987, p.457.
40. K. Kjær, J. Als-Nielsen, C.A. Helm, L.A. Laxhuber, and H. Möhwald, Phys. Rev. Lett. **58**, 2224 (1987).
41. J.H. Ipsen, G. Karlström, O.G. Mouritsen, H. Wennerström, and M.J. Zuckermann, Biochim. Biophys. Acta **905**, 162 (1987).
42. B. Michels, N. Fazel, and R. Cerf, Eur. Biophys. J. **17**, 187 (1989).
43. H. L. Scott and W. H. Cheng, Biophys. J. **28**, 117 (1979).
44. B.J. Alder and T.E. Wainwright, Phys. Rev. **127**, 359 (1962).
45. K.J. Strandburg, Rev. Mod. Phys. **60**, 161 (1988).
46. O. Albrecht, H. Gruler, and E. Sackmann, J. Physique **39**, 301 (1978).

# Part II

# Quantum Systems

# Numerical Evaluation of Candidate Wavefunctions for High-Temperature Superconductivity

*R. Joynt*

Department of Physics and Applied Superconductivity Center,
University of Wisconsin-Madison, Madison, WI53706, USA

Abstract. This article reviews the variational Monte Carlo approach to the Hubbard model and the problem of high-temperature superconductivity. An historical introduction to the conceptual background of the problem is given, and the basic technical details of the method are summarized. The chief physical results, particularly the prediction of antiferromagnetic fluctuations in almost localized Fermi liquids, are outlined. The consequences for high-temperature superconductivity are drawn, and future prospects are speculated upon.

## 1. Introduction

The Hubbard model has been for many years the paradigmatic model for interacting electron systems [1]. It is written in its simplest form for a single band of electrons as

$$H = \sum_{ks} \varepsilon_{ks} n_{ks} + u \sum_i n_{i\uparrow} n_{i\downarrow}.$$

$\varepsilon_{ks}$ are the band energies of the Bloch states labelled by k and the interaction is taken to act only within a unit cell labelled by i. The short range approximation for the interaction may be justified if there is a second set of electrons which gives rise to screening of the primary set; even if only one band is important, it may be possible to treat the long-range part of the Coulomb interaction in a mean-field fashion and incorporate the resulting renormalizations into the $\varepsilon_{ks}$. It may be that the major part of the problem of correlation would be understood if we could solve the Hubbard model. The fact that we can't (except in one dimension) supports this view!

In the 60's most of the work on this model focussed on questions of metallic magnetism, particularly ferromagnetism. Hartree-Fock treatment of the ground state of the model gives the Stoner criterion for ferromagnetism:

$$u\, D(\varepsilon_F) > 1, \text{ where } D(\varepsilon_F) = \sum_k \delta(\varepsilon_F - \varepsilon_k),$$

and $\varepsilon_F$ is the Fermi energy. This is very successful in a phenomenological sense, telling us why cobalt and nickel are ferromagnetic but titanium and scandium are not.

In a deeper sense, however, this situation is not very satisfactory. The Stoner criterion tells us that the interesting regime for U is the intermediate to strong coupling regime. It is evident that the Hartree-Fock treatment cannot be correct in this regime, since it is identical to first-order perturbation theory in U. This will very much overestimate the effect of U. In the actual ground state, opposite spin electrons will find ways to avoid one another, whereas in the unperturbed ground state they don't even try. This problem was somewhat alleviated by the work of Kanamori [2], who calculated how much U should be reduced in the intermediate regime. The extreme large-U limit was shown to be ferromagnetic in the limit of a half-filled band with a single-hole by Nagaoka [3], but the connection of this work to the previous work was not clear. The Nagaoka mechanism of ferromagnetism is that the kinetic energy of a hole moving through an ordered background of parallel moments is less than that of a hole moving through a disordered or antiferromagnetic background. This is totally different from the mean-field mechanism, which arises from minimizing the potential energy. It does not appear that the two regimes could ever be smoothly connected. Nor is it even clear that the Nagaoka result is valid over a finite range of the parameter space, [4], [5].

Parallel to the mean-field treatment was the work of Gutzwiller [6], which was a variational treatment. The ground state was taken to be

$$\Psi_G = e^{-\lambda d} \left|e^{i\vec{k}\cdot\vec{r}}\right|_\uparrow \left|e^{i\vec{k}\cdot\vec{r}}\right|_\downarrow, \text{ where } d = \sum_i n_{i\uparrow} n_{i\downarrow}$$

is the number of doubly occupied sites, and $\left|e^{i\vec{k}\cdot\vec{r}}\right|_\downarrow$ is a Slater determinant. $\lambda$ is a variational parameter which will clearly increase as U increases. This will take into account the effect mentioned above, namely that the electrons will avoid one another. Gutzwiller worked out clever approximate ways of evaluating matrix elements with his wavefunctions, and was able to determine $\lambda(U)$. However, since the chief interest was in ferromagnetism, and this method didn't give ferromagnetism for any U, it was not immediately followed up. The physical ideas involved clearly go beyond ferromagnetism, since this is a rather general way to handle correlation. It is the analog in lattice systems of so-called Jastrow wavefunctions in continuum systems. Brinkman and Rice made use of Gutzwiller's results for a quite different problem a few years later [7]. Think of a system with a large U. Furthermore, let the band be exactly half-filled. The idea is that there will come a point ($U = U_c$) where the system will become an insulator because the electrons will occupy the sites singly - the avoidance becomes total. Such a system can't conduct electricity because there is a gap for charge-carrying excitations. Brinkman and Rice pointed out that the Gutzwiller approximation did indeed have such a transition where $<d> \to 0$ at a finite U value, with the effective mass diverging at the same point. The object was to explain metal-insulator transitions in vanadium oxide and other systems. One can think of such transitions as localization transitions, or crystallization. Such transitions, when by interactions, are called Mott transitions. The Brinkman-Rice theory opens up the possibility of very large mass systems, because some systems, particularly those containing 4f electrons, probably have very large values of U. Close to half-filling we expect heavy electrons in this case, as long as the ferromagnetic instability is somehow avoided. The heavy fermion systems such as $CeAl_3$ seem to fall into this category. Why none of them are ferromagnetic (though many are antiferromagnetic) remains mysterious.

This was the situation obtaining in 1986, when the high-temperature superconductors were discovered. The Hubbard model seemed to be very rich, containing magnetism, the metal-insulator transition, and heavy electron behavior, all of which are found in nature. Hence the model still seems to retain its status as the paradigm for correlated electron systems. However, the interplay of these different phenomena is still, 25 years later, not well understood.

With this background, it was perhaps natural for Anderson to suggest that the Hubbard model also contained superconductivity [8].


2. Recent History

In 1982 Kaplan et al. [9] discovered an interesting connection between the Gutzwiller wavefunction for a half-filled band as $\lambda \to \infty$ (no doubly occupied sites) and the wavefunction for the ground state of the one-dimensional Heisenberg antiferromagnet. In one dimension, the Gutzwiller wavefunction had essentially the same nearest-neighbor spin correlation as Bethe's exact solution! These authors showed this by evaluating this quantity on finite one-dimensional chains. This was totally unexpected because the Gutzwiller wavefunction was designed not to enhance spin correlation but rather to minimize interaction energy. The spin correlations in the Slater determinants, i.e., the exchange hole, are of course antiferromagnetic. It appeared that these were vastly increased (by a factor of four) by applying the Gutzwiller procedure. In the extreme case of very large U, this procedure amounts to a projection. The original wavefunction is projected onto the subspace in which all sites are at most singly occupied. For a half-filled band, this means that the Gutzwiller wavefunction is a spin wavefunction. Horsch and Kaplan [10] later refined these calculations and inaugurated the use of the Monto Carlo method for carrying them out on larger systems.

The physical basis for the strong short-range correlation is the following. In a one-dimensional metal with a half-filled band, states with $|k| < \pi/2$ are occupied. On the average, such states will oscillate just so as to have maxima on every other site. If the projection operator is now applied, it forces an up spin and down spin electron to interweave their maxima, so that very strong antiferromagnetic correlations result.

The immediate utility of this connection between ordinary fermion determinants and spin systems is to provide a new language for speaking about spin models. If there is such strong correlation, then it should be possible to use these Gutzwiller wavefunctions for the solution of spin problems. As Haldane [11] and Shastry [12] showed independently, the one-dimensional wavefunction is indeed the exact ground state of a Heisenberg model with rather long-range interactions. It is probably possible to extend this result in some way to higher dimensions, but so far no one has succeeded in doing this.

There is, however, a deeper significance to the fact that these antiferromagnetic correlations are present. This has to do with the superexchange mechanism. In 1959 Anderson [13] showed that, in insulators, virtual transitions from one site to another with the intermediate state having a doubly occupied site will lead to an antiferromagnetic coupling. This explains why Mott insulators are so often antiferromagnetic. (Note that this work antedates the Hubbard model by several years). For the nearest neighbor hopping model,

$$H = -t\sum_{ij}( c_i^+ c_j + h.c. ) + U \sum_i n_{i\uparrow} n_{i\downarrow} ,$$

this means we can write an approximate Hamiltonian

$$H_{eff} = -t\sum_{ij}(c_i^+ c_j + h.c.) + J \sum_{ij}(\vec{S}_i \cdot \vec{S}_j - n_i n_j/4) ,$$

which is valid only in the low-energy subspace, that is the subspace in which configurations with doubly occupied sites do not appear. In these equations i,j must be nearest neighbors and each such pair is counted once. $J = 4t^2/U$. The point, first noted by Gros et al. [14], is that the Gutzwiller wavefunction is likely to be a very good candidate for the ground state also of metallic systems. Both parts of the energy, as reflected in the two terms of the effective Hamiltonian, should be reasonably well minimized. The kinetic energy should be low because the k-states in the Slater determinants are drawn from the original Fermi sea. The antiferromagnetic energy is good because these k-states are also ideal for the interweaving process mentioned above, at least near half filling. Gros et al. confirmed these speculations for the one-dimensional case. Specifically, they showed that

$$\langle \Psi_G | H_{eff} | \Psi_G \rangle = -2\alpha (1-n) t - 4\beta t^2 / U$$

with $\alpha \simeq 0.95$ and $\beta \simeq 0.1474$. Here n is the hole density and the results are for n much less than one. The absolute maximum for $\alpha$ is for the uniformly magnetized state and is equal to one. The absolute maximum for $\beta$ is surely given by the Bethe Ansatz value at half-filling and is equal to 0.1477. At least in one dimension the Gutzwiller wavefunction is certainly a very good approximation to the ground state. For cases where comparison could be made to the exact solution of Lieb and Wu [15], the energies were very close. The realization that the almost localized system contained very strong antiferromagnetic correlations checked out well with neutron scattering experiments being performed at that time on heavy fermion systems[16], at least on a qualitative level.

The method was generalized by Shiba[17], and Yokoyama and Shiba[18] for the periodic Anderson model, more appropriate for heavy fermion systems than the Hubbard model in important respects.

This work was all performed in one dimension, because of numerical limitations and the desire to compare to known results. There was good reason to suppose that the kinetic part of the energy of the ground state would continue to be favorable in higher dimensions, since the selection of k-values can still be arranged in the proper fashion. The spin part of the energy is another matter, however. In one dimension, $k_F$ is halfway to the zone boundary, which is the proper distance for the interweaving argument to hold. In two dimensions, $k_F$ is $1/\sqrt{2}$ of the way to the zone boundary, which is not quite so good. Also, the ground state in two dimensions was believed to have antiferromagnetic long-range order. The Gutzwiller wavefunction is probably worst at half filling and improves as holes are added to the system. Indeed, it was found that the energy of the wavefunction at half filling was -0.267 J per bond, whereas the ground state energy was known from exact diagonalization of small systems to be about -0.333 J.

This was roughly the situation in 1986.

## 3. High-Temperature Superconductors

The high transistion temperatures of the high-$T_c$ materials immediately suggested that the energy scale of the interaction responsible for superconductivity was high, perhaps as high as the scale of the electron-electron interaction. Together with the very large spatial separation of the Cu-O planes, this pointed the way towards using the two-dimensional Hubbard model for the energetics of high-$T_c$ [8].

Two things pointed away from using this model, however. One was the fact that the Cu-O planes have unit cells consisting of three atoms with several Cu and O bands near the Fermi energy. Another was simply that the interaction in the Hubbard model is repulsive, and superconductivity is supposed to require attractive interactions. Both of these points are to some extent still controversial, and a thorough discussion of them is beyond the scope of this paper. As to the first point, it is at least known that a consistent description of the Cu-O planes as a one-band Hubbard model is possible in a certain parameter range[19]. As to the second, we at least have the counterexample of superfluid $^3$He, though admittedly $T_c$ is around 1 mK for that system, which is hardly high-$T_c$. Whatever the fundamental justification, the two-dimensional Hubbard model has become by far the single most popular model for high-temperature superconductivity.

The first investigation of superconductivity using the variational Monte Carlo method was the paper of Gros et al.[20]. This paper was the analogue of the Cooper problem for the Hubbard model. The question it answered was whether, for the effective Hamiltonian $H_{eff}$, it is possible to lower the energy by constructing a coherent superposition of Gutzwiller wavefunctions, each containing a single pair of holes. The answer was very interesting. It turned out that this was possible only if the holes were paired in a d-wave state. The easiest way to see what this means is to think of the BCS wavefunction

$$\Psi_{BCS} = \Pi \, (u_k + v_k \, a_{k\uparrow} a_{-k\downarrow}) |0\rangle$$

and apply the projection operator to it, that is, only allow configurations without doubly occupied sites. If

$$\Delta(k) = v(k)/u(k) = \Delta(\cos k_x - \cos k_y)$$

then the Cooper pair wavefunction has d-like character, in the sense that it changes sign four times as k encircles the origin. In the original paper $\Delta$ was defined only at four points, but Gros[21] later found a method to explore the full projected BCS wavefunction. s-wave superconductivity, defined as $\Delta$ = constant, was also tried and any nonzero $\Delta$ of this form was always found to raise the energy. Remarkably, the inclusion of the d-wave correlations substantially improves the spin energy. At half filling, the energy of the wavefunction with $\Delta = 1.0\, t\, (\cos k_x - \cos k_y)$ is -0.318 J, as compared to -0.333 for the best antiferromagnetic wavefunctions. And the d-wave superconducting wavefunction does not have long-range magnetic order.

This was support for the "resonating valence bonds" ideas, whose basis was that one could combine spin singlet pairs to get a low energy but no Neel order. Very similar variational results on superconducting wavefunctions were obtained also by Yokoyama and Shiba[22] using a rather different method.

There have recently been some further generalizations of this work. Lee and Feng[23] wrote down a state which incorporates both a spin density wave and d-wave superconductivity. They multiply the BCS state by $e^{S/\hbar}$, where $S = \Sigma S_z(x,y)(-1)^{x+y}$ is the total staggered magnetization and h is a variational parameter. In this fashion they were able to lower the energy in the half-filled case to -0.332 J, and obtain correspondingly good results away from half filling for the metallic state. The antiferromagnetism appears to go away very quickly as holes are added, at a concentration of a few per cent. Compared with the experiments, this is too low, but the real system has disorder, and that may shift the phase boundary considerably.

One thing which is clearly missing in the Gutzwiller wavefunction is spin polaronic effects. These might be expected to be important because the Nagaoka theorem lends weight to the view that a hole should form a ferromagnetic cloud around itself. This was investigated by Coppersmith and Yu[24]. They put the extra hole-spin correlations into the wavefunctions by a multiplication factor similar to the one of Lee and Feng. Surprisingly, they found that the formation of such polarons always seems to raise the energy. This appears to be due to the fact, noted already in the one-dimensional case, that the kinetic energy of the hole is not as strongly affected by the total spin of the background as one would expect.

Finally one can try a completely different starting state for the original Hartree-Fock wavefunction to the right of the projection operator. Liang and Trivedi[25] used filled Landau levels rather than a filled Fermi sea, motivated by ideas about "flux phases". They were indeed able to improve on the spin correlation energy of the ordinary Gutzwiller state, obtaining an energy of -0.31 J in the half filled limit. Although still higher in energy than the other states mentioned above, these calculations open up strange and perhaps promising areas for exploration.

## 4. Some Recent Results

In this section I outline very briefly some preliminary results obtained in collaboration with G. -J. Chen, F. C. Zhang, and C. Gros. We have slightly improved the wavefunction of Lee and Feng by using Hartree-Fock spin density wave operators. We write

$$\Psi = \prod_k (u_k + v_k d^\dagger_{k\uparrow} d^\dagger_{-k\downarrow}) |0\rangle$$

where $d^\dagger_{ks} = \alpha_k a^\dagger_{ks} + \text{sgn}(s)\beta_k a^\dagger_{k+Qs}$. Here $Q = (\pi,\pi)$ is the wavevector of the antiferromagnetism. $\alpha_k, \beta_k$ are variational parameters which determine the amplitude of the staggered magnetization. We have mapped out the entire plane of n (=number of holes per site) and J/t. We find coexistence of d-wave superconductivity and antiferromagnetism up to about n = 0.05, in agreement with Lee and Feng. Much more surprisingly, we find that d-wave superconductivity is preferred over the normal state up to about n = 0.4 for a large range of J/t, roughly from 0.2 to 0.5. The calculation at such high hole densities requires some extension of $H_{eff}$ ( the inclusion of so-called three site terms), and this makes the calculations considerably more difficult. The result is surprising, since at such high doping levels there are many fewer spin bonds in the system and the superconductivity is always driven by the spin energy. However, this dividing line into the normal state is in good agreement with experiment, since the doping levels are high. In $YBa_2Cu_3O_7$, n is nominally 0.33.

We have done an extensive search of the "extended s" and d-wave parameter space. That is, we use a two parameter gap function

$$\Delta(k) = (\Delta_s + \Delta_d) \cos k_x + (\Delta_s - \Delta_d) \cos k_y .$$

The relative phase of $\Delta_s$ and $\Delta_d$ is taken to be either pure imaginary (s + id state) or pure real (s ∓ d state). In all cases we find that the pure d-wave state is always the lowest in energy. The last conclusion is not in good agreement with experiment at present. Recent photoemission measurements of the k-dependence of $\Delta$ seem to show that it is roughly isotropic at low temperatures[26]. But there is a long way to go in both theory and experiment on this question.

## 5. Future Directions

The variational Monte Carlo seems to be the most reliable way to check out theories of the ground state of the large-U Hubbard model at the present time. Unfortunately, very few measurable properties depend only on the ground state wavefunction. One very important one is the frequency-integrated neutron scattering cross section at low temperature. This is just an equal-time correlation function of just the sort calculable by this method. We are presently computing this quantity for various wavefunctions. To get the frequency dependence we would need the properties of excited states – this is beyond the capabilities of the method at the present time and is perhaps the single most pressing problem on the technical side.

On the side of physics, we need answers to other questions. Is the Hubbard model all one really needs ? Terms left out may be small, but so is the interaction term in the BCS Hamiltonian. Do we really understand the normal state ? There are many indications that it is not a Fermi,liquid, which would cut out the Gutzwiller state. Are the flux phases really the answer above $T_c$ ?

Both as a check on analytic theories and as a way to investigate new ideas which have no good analytic treatment, variational Monte Carlo is sure to be in on the kill on the high-$T_c$ problem.

## 6. Acknowledgements

The above is the result of much fruitful interaction with G. J. Chen, F. C. Zhang, C. Gros, and T.M. Rice, but the mistakes are mine alone.

## 7. References.

[1] J. Hubbard, Proc. Roy. Soc. A276, 238 (1964)
[2] J. Kanamori, Prog. Theor. Phys. 30, 275 (1963)
[3] Y. Nagaoka, Phys. Rev. 147, 392 (1966)
[4] M. Takahashi, J. Phys. Soc. Japan 51, 3425 (1982)
[5] Y. Fang, A. Ruckenstein, E. Dagotto, and S. Schmitt-Rink. Phys. Rev. B40, 7406 (1989)
[6] M. C. Gutzwiller, Phys. Rev. 137, A1726 (1965)
[7] W. F. Brinkman and T. M. Rice, Phys. Rev. B2, 4302 (1970)
[8] P. Anderson, Science 235, 1196 (1987)
[9] T. A. Kaplan, P. Horsch, and P. Fulde, Phys. Rev. Lett., 49, 889 (1982)
[10] P. Horsch and T. A. Kaplan, J. Phys, C 16, L1203 (1983)
[11] D. Haldane, Phys. Rev. Lett. 60, 635 (1988)
[12] B. Shastry, Phys. Rev. Lett. 60, 639 (1988)
[13] P. W. Anderson, Phys. Rev. 115, 2 (1959)
[14] C. Gros, R. Joynt, and T. M. Rice, Phys. Rev. B36, 381 (1970)
[15] E. H. Lieb and F. Y. Wu, Phys. Rev. Lett. 20, 1665 (1968)
[16] G. Aeppli, A. Goldman, G. Shirane, E. Bucher, and M. Lux-Steiner, Phys. Rev. Lett. 58, 808 (1987)
[17] H. Shiba, J. Phys. Soc. Japan 55, 2765 (1986)
[18] H. Yokoyama and H. Shiba J. Phys. Soc. Japan 56, 3570 (1987)
[19] F. C. Zhang and T. M. Rice, Phys. Rev. B37, 3759 (1988)
[20] C. Gros, R. Joynt, and T. M. Rice, Z. Phys. B68, 425 (1987)

[21] C. Gros, Phys. Rev. B38, 931 (1988)
[22] H. Yokoyama and H. Shiba, J. Phys. Soc. Japan 57, 2482 (1988)
[23] T. K. Lee and S. Feng, Phys. Rev. B38, 11809 (1988)
[24] S. Coppersmith and C. Yu, Phys. Rev. B39, 11464 (1989)
[25] S. Liang and N. Trivedi, Phys. Rev. Lett. 64, 232 (1989)
[26] C. G. Olson, (private communication)

# Two-Dimensional Quantum Antiferromagnet at Low Temperatures

*E. Manousakis*

Department of Physics and Center for Materials Research and Technology,
Florida State University, Tallahassee, FL 32306, USA

We discuss the spin-dynamics in the spin-$\frac{1}{2}$ antiferromagnetic Heisenberg model and the related quantum nonlinear $\sigma$ model on the square-lattice at zero and low temperatures. We use mainly numerical methods, including variational Monte Carlo, quantum Monte Carlo and exact numerical diagonalization techniques. We compare our results to those obtained by other techniques and to certain experimental data obtained on undoped copper-oxygen antiferromagnetic insulators. Effects of hole-doping are briefly summarized in the framework of the strong-coupling Hubbard model.

## 1. Introduction

The Hubbard model in the large Coulomb-repulsion limit is under intensive theoretical study because it may be relevant to the physics behind the copper-oxide superconductors. Regardless of its connection to the high-temperature superconductors, the physics of a strongly-correlated electronic system should be understood at least in its simplest theoretical abstraction: the single-band Hubbard model. Low-order strong-coupling perturbation treatment of the Hubbard model produces an effective Hamiltonian[1] which operates in a restricted Hilbert space having states with no double-occupancy below half-filling. Ignoring a three-site interaction term this effective Hamiltonian is now known as the $t - J$ model

$$H_{t-J} = -t \sum_{<i,j>,\sigma} (c^\dagger_{i,\sigma} c_{j,\sigma} + h.c) + J \sum_{<i,j>} \left( S^z_i S^z_j + \frac{1}{2}(S^+_i S^-_j + S^-_i S^+_j) \right), \qquad (1.1)$$

where $S^z_i = \frac{1}{2}(c^\dagger_{i\uparrow} c_{i\uparrow} - c^\dagger_{i\downarrow} c_{i\downarrow})$, $S^+_i = c^\dagger_{i\uparrow} c_{i\downarrow}$ and $S^-_i = c^\dagger_{i\downarrow} c_{i\uparrow}$ and $c^\dagger$ is the fermion creation operator in the Hubbard model. This Hamiltonian is assumed to operate only in a subspace of the Hilbert space with states having no doubly occupied site. The $t - J$ model is interesting on its own and its derivation from the Hubbard model may serve as a motivation. At half-filling the hopping term of this Hamiltonian is inactive and thus (1.1) reduces to the spin-$\frac{1}{2}$ antiferromagnetic (AF) Heisenberg model (AFHM) on the square-lattice

$$H = J \sum_{<ij>} \vec{S}_i \cdot \vec{S}_j. \qquad (1.2)$$

The magnitude of the characteristic energy scale $J$ of the model may be determined from the magnetic properties of the undoped materials. In two space dimensions (2D) the Heisenberg model cannot develop long-range-order (LRO) at any non-zero temperature[2]. Doubts expressed about the existence of LRO in the ground-state, because of the low-spin case and low-dimensionality of the model, are now more or less quenched. Even though there is no rigorous proof of the existence of AF LRO in the ground-state of the spin-$\frac{1}{2}$ 2D antiferromagnet[3], the picture of spin-wave theory[4] (SWT) has recently received significant support by several systematic approaches[5-10].

In the first part of this paper, we study the ground-state of (1.2) using the variational technique. We obtain[9,10] a variational wave function consistent with sum-rules for the dynamic structure function which gives accurate ground-state properties. Using this wave function, the sum-rules and the variational Monte Carlo (VMC) technique, we determine the spin-wave velocity, the spin-stiffness constant and staggered magnetization. The spin-wave velocity, the sublattice magnetization and the AF coupling $J$ are accessible to experiments and this allows a direct test of the theory.

In section 3 of this paper we study the model (1.2) at finite temperatures using quantum Monte Carlo methods[11,12]. We also study the quantum nonlinear sigma model (QNL$\sigma$M)[13] which is believed to be equivalent to the AFHM at low $T$[14]. The idea that models such as (1.1) may be relevant for the physics behind these materials is supported by the comparison of the temperature dependent spin-spin correlation length $\xi(T)$ obtained from these studies with that inferred by neutron scattering experiments[15]. We find that for the model (1.2), $\xi(T \to 0) = a exp(2\pi\rho_s/T)$, and the same form is valid for the QNL$\sigma$M in the regime controlled by its ordered phase and at low $T$. One- and two-loop Renormalization Group calculations for the sigma model have been carried out by Chakravarty, Halperin and Nelson (CHN)[16,17]. Our results disagree with the results of their one-loop calculation[16] and are in agreement with their results obtained by calculating the $\beta$-function up to two-loop order[17]. Furthermore, our $\xi(T)$ agrees reasonably well with the neutron scattering data using $J = 1480K$. We conclude that SWT correctly describes the $T = 0$ properties of the spin-$\frac{1}{2}$ AFHM and the 2D thermal spin correlations in the model are consistent with those observed by neutron scattering.

Due to the fact that the Monte Carlo simulations of the model (1.1) are hindered by problems arising from the fermion statistics, the information about the phase diagram and superconductivity in the model is limited. In section 4 of this paper, we report certain results for the single-hole band obtained by VMC methods[18]. A wave function which takes into account spin-spin and spin-hole correlations is constructed following the ideas of Feynman-Cohen for the case of a helium-three atom moving in a background of liquid helium-four. Using this wave function and the VMC technique we calculate the hole excitation spectrum. For the case of two or more holes exact diagonalization studies[19] on small-size system indicate that there is a range of $t/U$ where hole-pairing is possible.

## 2. Zero Temperature Properties

Antiferromagnets were initially treated using spin-wave theory[4] which assumes AF LRO in the ground-state and treats the zero-point motion of small quantum fluctuations about the classical Néel state perturbatively. This approach for a spin-$S$ antiferromagnet is an expansion in powers of $1/(zS)$, $z$ being the coordination number. The role of quantum fluctuations becomes more important for small $S$, and low-dimensional systems and it is natural to raise doubts about the convergence of this approach for the smallest possible spin case and for 2D spin-systems.

There is no rigorous proof of the existence or non-existence of AF LRO in the ground-state of (1.2) on the square-lattice for spin-$\frac{1}{2}$. However, a growing number of calculations[5-10] suggest that the picture obtained from SWT is correct. For example, using the Green's function Monte Carlo (GFMC) method, Carlson and independently Trivedi and Ceperley have performed accurate simulations[8] on up to $32 \times 32$ lattice size. They find that the ground-state energy per bond and staggered magnetization follows the finite-size scaling expected from SWT. Their extrapolated values to the infinite lattice are $E_0/NJ = -0.6692 \pm 0.0002$ and $m^\dagger \simeq 0.31 \pm 0.01$. These results compare very well with those obtained from SWT[4], which are, respectively, 0.303 and $-0.6704$. Furthermore, they studied spin-wave states and concluded that the energy-gap scales as predicted by SWT and vanishes in the infinite-lattice limit.

In this section, we shall review certain results obtained with the variational approach. In Ref. 9, using a complete set of multi-magnon states, we calculated the matrix elements of the Hamiltonian (1.2) in a separability approximation originally developed for the treatment of strongly correlated quantum liquids. The ground-state wave function in this approximation is obtained as

$$|\psi_0> = \exp\left(-\frac{1}{2}\sum_{i<j} u_{ij}\sigma_i^z\sigma_j^z\right)|\phi>, \qquad (2.1.a)$$

$$u_{ij} \equiv \frac{1}{N}\sum_{\vec{k}}\left(\sqrt{\frac{1+\gamma(k)}{1-\gamma(k)}} - 1\right)e^{i\vec{k}\cdot(\vec{R}_i-\vec{R}_j)}, \qquad (2.1.b)$$

where $\gamma(\vec{k}) = 1/2(cos(k_x) + cos(k_y))$ and

$$|\phi> \equiv \frac{1}{\sqrt{2^N}}\sum_c (-1)^{L(c)}|c>. \qquad (2.1.c)$$

The sum is over all possible spin configurations $c$ of the lattice and $L(c)$ is the number of down spins in one sublattice contained in the configuration $c$. Therefore the state $|\phi>$ can be written as $|\phi> = \prod_{\vec{R}\epsilon A}|\vec{R},+> \prod_{\vec{R}\epsilon B}|\vec{R},->$ and the states $|\vec{R},+>$ and $|\vec{R},->$ are the eigenstates of $\hat{S}_{\vec{R}}^x$. $A$ and $B$ represent the two sublattices. Hence the state (2.1.c) has zero staggered magnetization in the $z$ and $y$ directions but has full staggered magnetization in the $x$ direction. In fact, if we rotate the Néel state around the $y$-axis by $\pi/2$ we will obtain the state (2.1.c).

Variational wave functions of similar form have been studied earlier[20] by Hulthen and Kastelijn for one dimension and for one, two and three dimensions by Marshall and Bartkowski. More recently the same form was studied by Huse and Elser (HE) and by Horsch and Linden (HL) [21] using the VMC approach. HE took $u(1) = u_1$ and $u(r) = a/r^b$ for $r > 1$, where $r = |\vec{R}_i - \vec{R}_j|$ and treated $u_1$, $a$ and $b$ as variational parameters. The energy per site obtained in their calculation is $E_0/N \sim -0.664J$ for $u_1 \sim 0.65$, $a \sim 0.475$, and $b \sim 0.7$. HL used only $u(1)$ as a variational parameter (and $u(r > 1) = 0$) and they found $E_0/N \sim -0.644J$. Notice that our $u$ is not a function of the distance $r$ between two points on the lattice but rather a function of the two components $x$ and $y$ of the vector $\vec{R}_{ij}$. The form (2.1.b) has long-distance behavior consistent with the existence of long-wavelength spin-wave excitations. From Eq. (2.1.b) we find that

$$u(r \to \infty) = \frac{\sqrt{2}}{\pi r}. \qquad (2.2)$$

The tails of the wave function of Ref. 21 and that of Eq. (2.2) are different. We have calculated the expectation value of the Heisenberg Hamiltonian (1.2) with the wave function (2.1) using MC integration and restricting the sum in $|\phi>$ (Eq. 2.1.c) over configurations having zero total $S_z$ only. We found almost the same energy (slightly better (lower)) as that of HE. The advantage of the wave function (2.1) is its simple physical origin and the fact that it gives the same ground-state energy with no free parameter. Next, we improve the wave function (2.1) further using sum-rules of the dynamical structure function. We also propose a method to calculate the spin-wave velocity accurately.

We can use a representation in which the eigenstates of the spin-$\frac{1}{2}$ AFHM are expressed as a superposition of configurations $|c>$ labeled by the locations of one kind of spins (say the down spins) on the lattice, i.e., $|c> = |i_1, i_2, ...i_r>$. In this representation, we can show that the eigenvalue problem reduces to a difference equation identical to the many-particle Schrödinger equation on a 2D lattice (the "particles" correspond to down spins). In this quantum lattice-gas of bosons, the "particles" have "mass" $m = 2$ (in units of $J$ and lattice-spacing $a$) and interact via a pair potential $V_{ij}$ having an infinite on-site repulsion, $V_{ij} = 1$ if $ij$ are n.n, otherwise $V_{ij} = 0$. This is a useful representation because our knowledge about the system of Bose-particles can be applied to the magnetic system also. For example, it is known that the ground-state of this system has a broken symmetry (condensate) which in the magnetic language corresponds to AF-LRO. The elementary excitations in the Bose-system are density fluctuations (phonons in the long-wavelength limit) which in the magnetic system correspond to spin-waves. Reatto and Chester have shown[22] that the zero-point motion of the long-wavelength modes of the Bose-system (zero-sound) gives rise to a long-range tail in the Jastrow wave function. For a 2D system, we obtain

$$u(r \to \infty) = \frac{mc}{4\rho_0 \pi r}, \quad (2.3)$$

where $c$ is the spin-wave velocity. The ground-state of the Heisenberg antiferromagnet has zero total $S_z$ and the number of down spins is exactly half the total number of sites giving $\rho_0 = 1/2$. Comparing the tails (2.2) and (2.3) we find $c = \sqrt{2}$ which is the value found by linear SWT.

We use[10] three $\omega$-moments (sum rules) of the spin-dynamical structure function $S(\vec{q}, \omega)$ to determine the spin-wave velocity, assuming that a single-magnon state exhausts them in the long-wavelength limit. Using the $\omega^0$-moment (the static structure function $S(\vec{q}) \equiv \langle 0|S^z_{-q} S^z_q|0\rangle$ ) and $\omega^1$-moment we obtain

$$c = \frac{f}{s_1}, \quad (2.4.a)$$

where $s_1$ is the slope of $S(\vec{q})$ and

$$f \equiv -\frac{J}{4} \sum_\delta \langle 0|(S^+_i S^-_{i+\delta} + S^-_i S^+_{i+\delta})|0\rangle. \quad (2.4.b)$$

The value of $c$ calculated from (2.4) is sensitive to both the tail of the wave function and finite-size effects. As in the case of quantum liquids we shall use a different and more accurate way to determine $c$, explained next.

We have obtained a third sum rule analogous to the compressibility sum rule in the case of quantum fluids which in the spin-system is translated to "magnetic susceptibility sum-rule" ($\omega^{-1}$ moment). Again assuming that this sum-rule in the long-wavelength limit is exhausted by a single-magnon-excitation we obtain

$$c = \sqrt{2f\epsilon''}. \quad (2.5)$$

Here $\epsilon''$ is the second derivative of the energy per particle $\epsilon(M)$ with respect to the magnetization density $M = 1/N < \sum_i S^z_i >$. We note that the magnetization density corresponds to the particle density in the Bose system and the energy $\epsilon(M)$ to the ground-state equation of state. We calculate $\epsilon''$ by restricting ourselves to a subspace with well-defined $S^z_{tot} = M$, i.e., total z-component of the magnetization. Therefore, we can determine the spin-wave velocity in a way analogous to that used in the case of quantum liquids to calculate the

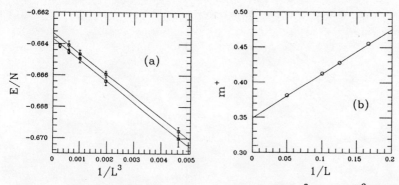

Figure 1. a) The energy per site as a function of $L^{-3}$ for an $L^2$ lattice. The results are obtained with the parameter-free wave function (2.1) (upper line) and with the improved wave function (lower line). b) The staggered magnetization versus $L^{-1}$ with the improved wave function.

sound velocity. This technique is known to be accurate for numerical studies, because it is not too sensitive to finite-size effects and to the tail of the ground-state wave function.

In the variational calculation we used the form (2.1.a), including in the sum (2.1.c) only states with zero magnetization, and took $u(1)$ and $u(\sqrt{2})$ as variational parameters and

$$u(\vec{r}) = \alpha u_0(\vec{r}), \quad for \quad \sqrt{x^2 + y^2} \geq 2, \qquad (2.6)$$

where $u_0(\vec{r})$ is that given by Eq. (2.1.b) and $\alpha$ is a parameter of order 1. We did not treat $\alpha$ as a variational parameter because the ground-state energy is not too sensitive to its precise value. Instead it is determined self-consistently by satisfying the third sum-rule (2.5). Given a value of $\alpha$, we perform the variational calculation and determine $\epsilon(M)$ and the spin-wave velocity $c$ from the slope of $\epsilon(M)$ via (2.5). Using the Chester and Reatto [22] relation (2.3), we obtain a new value of $\alpha$ from the relation $\alpha = c/\sqrt{2}$. This can be iterated until the input and the output value of $\alpha$ are the same. This procedure converges very quickly since as mentioned the energy $\epsilon(M)$ is not sensitive to the $\alpha$. Starting from $\alpha = 1$ the output value of $\alpha$ obtained from $\epsilon(M)$ is $\sim 1.2$. In the next step using the new value of $\alpha$ as input we obtain practically the same output value from $\epsilon(M)$.

The calculation is performed on lattices of several sizes up to $20^2$. In Fig. 1.a we present the ground-state energy per site as function of $L^{-3}$ [23,7] for lattices of size $N = L^2$. The energy obtained with the wave function (2.1.c) (upper line) and with the improved wave function (2.6) (lower line) are the same within error bars. The advantage of the improved wave function, however, is that it is consistent with the sum rules and gives a more accurate excitation spectrum. The extrapolated value of the energy upper bound to the infinite-size lattice is $-0.6637J \pm 0.0002$, while the best estimate for the exact value obtained by GFMC calculations[8] is $-0.6692J \pm 0.0001$. Fig. 1.b shows the square root of the expectation value of the square of the staggered magnetization $m^\dagger$ obtained with the improved wave function. We obtain $m^\dagger = 0.349 \pm 0.002$ for the infinite lattice. This is close to the values of $0.34 \pm 0.01$ and $0.31 \pm 0.01$ reported in Ref. 8, while their guiding trial functions give somewhat higher value of $m^\dagger$ than ours. In Fig. 2, we give $\epsilon(M)$ versus $M^2$ for several lattice-sizes. Notice that $\epsilon''$ is independent of the lattice-size within error bars and we obtain $\alpha = 1.22 \pm 0.02$. The value of this parameter (commonly called $Z_c$) is in good agreement with the value 1.158 obtained by spin-wave theory[4] and that obtained by GFMC [8] calculations and series expansions [24]. The same value of $\epsilon''$ is required to fit at small $M$ the exact diagonalization results for a $4 \times 4$ size lattice which have been obtained with methods explained in Ref. 19.

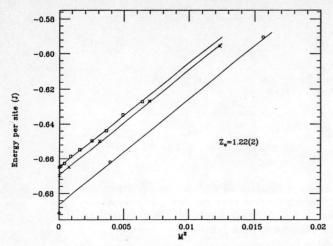

*Figure 2. The energy per site as a function of the square of the total magnetization $M^2$ for $4 \times 4$, $6 \times 6$, and $10 \times 10$ lattices. The slope is related to the spin-wave velocity via Eq. 2.5.*

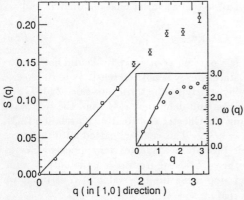

*Figure 3. $S(\vec{q})$ calculated with the wave function (2.6) with $\alpha = 1.22$. The straight line has a slope $s_1 = f/c$. The inset is the calculated $\omega(\vec{q})$. The straight line in the inset has a slope equal to $c = 1.22\sqrt{2}$ as obtained from (2.5)*

Hence, we expect the error in the calculation of $c$ due to the approximate nature of our wave function to be small.

In Fig. 3, we give the calculated $S(\vec{q})$ using $\alpha = 1.22$ in the wave function. The straight line has a slope $s_1 = f/c$. The inset is the $\omega(\vec{q})$ calculated assuming that a single-magnon excitation exhausts the $\omega^0$ and $\omega^1$ sum-rules. The straight line in the inset has a slope $c = 1.22\sqrt{2}$. Hence the spin-wave spectrum in the long-wavelength limit, although noisy as expected, is consistent with this sum-rule within error bars also. In our units $Z_\chi = 8\chi J = 8J/\epsilon'' = 0.667 \pm 0.004$, which is higher than the spin-wave value of 0.449 and close to that reported in Ref. 25.

## 3. Low Temperature Properties. Nonlinear $\sigma$ Model

Using Handscomb's quantum Monte Carlo method we have simulated[11] the spin-$\frac{1}{2}$ AFHM on the square lattice at finite temperatures. From the spin-spin correlation function we extracted the correlation length $\xi(T)$. We found that $\xi(T)$ increases very rapidly with decreasing temperature. In our first paper[11] we attempted to fit the temperature dependence of $\xi(T)$ to the form

$$\xi(T \to 0) = \frac{a}{T} exp\left(\frac{2\pi\rho_s}{T}\right), \qquad (3.1)$$

as suggested by spin-wave theory. We concluded that our numerical results are inconsistent with this form. We also attempted to fit our results to a Kosterlitz-Thouless form and obtained a good quality fit. In Fig. 4.a we reproduce our Fig. 3 from Ref. 11 for easy reference. The dashed-dotted line was our best fit of the numerical results (points with the error bars for various size lattices as indicated on the figure) to the form (3.1), while the solid line was the fit to the Kosterlitz-Thouless form. The dashed-line is our results of the leading contribution at high temperatures [12]. We suggested that either (a) the form (3.1) is incorrect or valid at lower temperatures inaccessible to our simulation technique or (b) topological excitations may play an important role in the thermodynamics of (1.2). The form (3.1) is also obtained by Chakravarty, Halperin and Nelson (CHN)[16], studying the equivalent QNL$\sigma$M by Renormalization Group with the $\beta$-function calculated with weak-coupling perturbation theory up to one-loop order and by the Schwinger-boson mean field calculations of Arovas and Auerbach and by the modified spin-wave theory by Takahashi at finite temperatures[26].

The improved two-loop calculation of CHN [17], however, shows that the prefactor in (3.1) is different, namely the correlation length at low $T$ behaves as

$$\xi(T \to 0) = C_\xi exp\left(\frac{2\pi\rho_s}{T}\right). \qquad (3.2)$$

Hence, CHN [17] point out that higher-order corrections are important and one should expect a constant prefactor. In Fig. 4.b we plot $Tln\xi(T)$ versus $T/J$ to demonstrate that (3.2) is a very good approximation to the Monte Carlo results of Ref. 11. The fit gives $\rho_s \simeq 0.22J$ and $C_\xi = 0.25$. This value of $\rho_s$ is in good agreement with the value found in section 2 by studying the ground-state of (1.2). On the basis of our results given in the

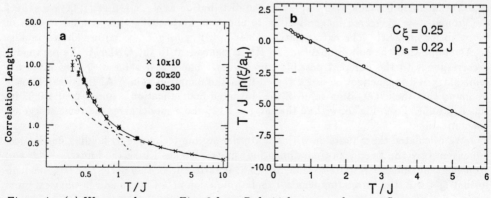

Figure 4. (a) We reproduce our Fig. 3 from Ref. 11 for easy reference. See text for explanations. (b) We plot the same data as $Tln\xi(T)$ versus $T/J$ to demonstrate that (3.2) is a very good approximation.

previous section, the results of this fit to the temperature dependence of $\xi$ and the results of our calculation of the $\sigma$ model explained next, we conclude that, even though a Kosterlitz-Thouless phase cannot be theoretically excluded, there is no need to invoke the presence of such transition. Furthermore, our calculations[11,12,13] strongly support the results of the two-loop calculation of CHN [17] and the form (3.2) at low $T$.

Next, we briefly outline the results obtained[13] on the quantum non-linear $\sigma$ model using both MC simulation and saddle point approximation. We shall show that the spin-$\frac{1}{2}$ AFHM corresponds to the ordered phase of the $\sigma$ model where $\xi(T)$ obtained from the $\sigma$ model in that phase agrees with the form (3.2).

The action for the nonlinear $\sigma$ model in two-space one-Euclidean time dimensions is defined as[14]

$$S_\sigma = \frac{\rho_\sigma}{2\hbar c_\sigma} \int_0^{\beta \hbar c_\sigma} d\tau \int dx dy \left( (\partial_x \vec{n})^2 + (\partial_y \vec{n})^2 + (\partial_\tau \vec{n})^2 \right). \tag{3.3}$$

Here $\vec{n}$ is a three-component vector field living on a unit sphere, $c$ is the spin-wave velocity and $\beta = 1/K_B T$. It is known that the model (3.3) can be derived[14] from the model (1.2) in the long-wavelength limit. The Euclidean-time is a result of the quantum nature of the problem, and is introduced via the Trotter-Suzuki formula in the calculation of the trace of $exp(-\beta \hat{H})$, when the operator (1.2) is used. The Heisenberg antiferromagnet maps[14] onto this same model for arbitrary spin in 2D, the different values of $S$ corresponding to different values of $\rho_s$ and $c_\sigma$. The field $\vec{n}(\vec{x})$ is proportional to the average staggered magnetization in the model (1.2). We discretize the space-time and put the model on the 2+1 dimensional lattice:

$$S_\sigma = -\frac{1}{2g} \sum_{\vec{x}} \sum_{\mu=1}^{3} \vec{n}_l(\vec{x}) \cdot \left( \vec{n}_l(\vec{x} + \hat{e}_\mu) + \vec{n}_l(\vec{x} - \hat{e}_\mu) \right), \tag{3.4}$$

where $g = \hbar c_\sigma/\rho_\sigma a$ and $\vec{x}$ covers the 2+1 dimensional lattice of lattice spacing $a$, size $L^2 L_\beta$ and

$$\beta \hbar c_\sigma = L_\beta a. \tag{3.5}$$

Let us first study the $T = 0$ case. In this case we consider $L_\beta = L \to \infty$ and the theory has only one parameter, the dimensionless parameter $g$. We have calculated the staggered magnetization expectation value defined as $\bar{n}_l = < \left( \frac{1}{L^3} \sum_{x_\mu} \vec{n}_l(x_\mu) \right)^2 >$, where the expectation value is taken with respect to the distribution $exp[-S_\sigma(\{\vec{n}_l(x_\mu)\})]$. We expect the ground-state staggered magnetization to obey the finite-size scaling similar to that of Fig. 1.b for $m^\dagger$[23,7]. The extrapolated values $n_0(g) \equiv \bar{n}_l(g, L \to \infty)$ are shown in Fig. 5. Assuming that the spin-S quantum AFHM is equivalent to the QNL$\sigma$M, the parameter $g$ corresponds to the different possible spin values. There is a value of $g$ that gives the same staggered magnetization as extracted from the quantum spin-$\frac{1}{2}$ AFHM. In section 2 we found that the expectation value of the staggered magnetization is about $0.60 \pm 0.02$ its classical value. From Fig. 5, we find that at $g = 1.125$ the $\sigma$ model gives the same staggered magnetization.

We calculated the $\beta$-function with a finite-size scaling technique by holding one lattice-dimension (say the Euclidean-time dimension) finite. We shall take the limit $L \to \infty$ and keep the time dimension finite so that Eq. (3.5) is satisfied. If, therefore, $L$ is large enough so that $\xi_l << L$, the correlation length is only a function of $L_\beta$ and $g$ and in physical units is given by

$$\xi = \xi_l(g, L_\beta) a. \tag{3.6}$$

Substituting $a$ from (3.5) in (3.6) we obtain

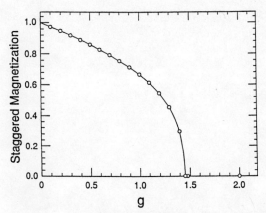

Figure 5. The extrapolated values $n_0(g) \equiv \bar{n}_l(g, L \to \infty)$.

$$\frac{\xi}{\beta \hbar c_\sigma} = \frac{\xi_l(g, L_\beta)}{L_\beta}. \tag{3.7}$$

At this point we wish to approach the continuum limit and hold the correlation length $\xi$ fixed in units of the length-scale $\hbar c_\sigma \beta$, while $L_\beta \to \infty$. Clearly this can be done provided that a value of $g = g_c$ exists at which $\xi_l(g_c, L_\beta \to \infty) \to \infty$ in such a way that the ratio

$$b = \frac{\xi_l(g, L_\beta)}{L_\beta} \tag{3.8}$$

remains constant. A similar discussion in the framework of finite-size scaling at the phase transition can be found in a paper by Brèzin[27]. The critical point $g = g_c$ is a fixed point of the scale transformation $L_\beta \to L'_\beta$ and $g \to g'$ defined as follows:

$$\frac{\xi_l(g, L_\beta)}{L_\beta} = \frac{\xi_l(g', L'_\beta)}{L'_\beta}. \tag{3.9}$$

For large $L_\beta$ this equation defines $g(L_\beta)$, which via (3.9) gives the function $g(a)$. In Fig. 6, $b$ is given as a function of $g$ calculated at $L_\beta = 2, 4, 8$ (for large enough space extent $L$ so that finite-size effects on $\xi_l$ due to finite $L$ are negligible). We clearly see the presence of a critical point at $g_c \simeq 1.45$. Compare this value of $g_c$ and the value of $g_c$ where $n_0(g)$ vanishes in Fig. 5. The $\beta$ function (and $a(g)$) determined via (3.9) close to the critical point is given by

$$\beta(g) = -\beta_1(g - g_c) + ... \tag{3.10.a}$$

with $\beta_1 \simeq 1.3 \pm 0.1$. By integrating the equation $\beta(g) = -a \frac{dg(a)}{da}$ we obtain

$$a(g) = a_0 |g - g_c|^\nu \tag{3.10.b}$$

where $\nu = 1/\beta_1 \simeq 0.77 \pm 0.07$, while the textbook value for $\nu$ is $\sim 0.7$. The results do not depend strongly on the errors in the determination of the function $a(g)$, i.e., the precise value of $\nu$. Here, using $\nu = 0.7$, we find essentially the same results as those obtained in Ref. 13.

Having approximately determined the function $a = a(g)$, we can proceed to study the behavior of the correlation at low $T$. Using Eq. (3.5) and $a(g)$ we find that

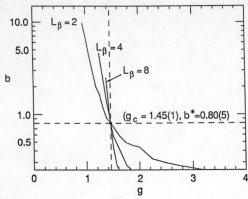

Figure 6. The ratio $b = \xi_{\text{latt}}/L_\beta$ versus $g$ for different $L_\beta$. Notice that all the lines for different $L_\beta$ pass through the same point $(g_c, b^*)$.

$$t \equiv \frac{T}{T_0} = \frac{1}{L_\beta |g - g_c|^\nu}, \qquad (3.11.a)$$

where the constant temperature scale $T_0$ is defined as

$$k_B T_0 \equiv \frac{\hbar c_\sigma}{a_0}. \qquad (3.11.b)$$

Furthermore, substituting (3.10.b) in (3.6) we obtain

$$\xi_0 \equiv \frac{\xi}{a_0} = \xi_l(g, L_\beta)|g - g_c|^\nu. \qquad (3.11.c)$$

Namely, given the value of the correlation length $\xi_l(g, L_\beta)$ for given values of $g$ and $L_\beta$, using the Eqs (3.11) we can calculate the physical correlation length in units of the constant length-scale $a_0$ and the physical temperature in units of the constant temperature-scale $T_0$ defined by (3.11.b). Using the calculated correlation lengths for various values of $g$ and $L_\beta$ we calculate $\xi_0$ and the corresponding $t$ (using (3.11.a) and (3.11.c)) and plot them to obtain a single curve $\xi_0(t)$, shown in Fig. 7.a.

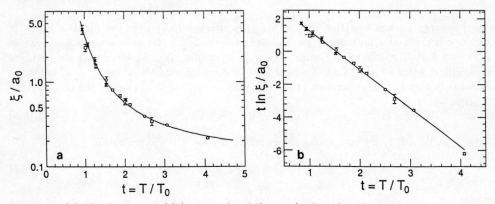

Figure 7. (a) The function $\xi_0(t)$ (see text for definition). Our data for various $g$'s collapse on the same curve by using the calculated renormalization group $\beta$-function. The solid line corresponds to an exponential fit. (b) Demonstration that the correlation length as a function of $T$ can be approximated by $ae^{b/T}$.

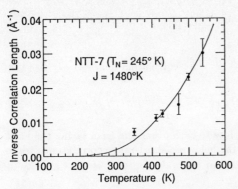

*Figure 8.* The solid line corresponds to an exponential fit to both our results for the nonlinear $\sigma$ model and the spin-$\frac{1}{2}$ AFHM. We used the value $J = 1480K$ for the AF coupling. The solid circles with error bars are neutron scattering data taken on the insulator $La_2CuO_4$.

The fact that $\xi(T)$ obeys the expression obtained by the two-loop calculation of CHN, i.e., Eq. (3.2), is demonstrated in Fig. 7.b where the function $tln(\xi_0(t))$ can be approximated by a straight line. A straight-line fit gives

$$\xi/a_0 = A exp\left(B\frac{T_0}{T}\right) \qquad (3.12)$$

with $A = 0.093 \pm 0.002$ and $B = 3.64 \pm 0.02$. The constants $a_0$ and $T_0$ cannot be determined within the non-linear $\sigma$ model since there are two parameters in the model, namely $g$ and $c_\sigma$. Assuming that the $\sigma$ model describes the continuum limit of the $S = 1/2$ AFHM, we can determine $a_0$ and $T_0$ in terms of $J$ and the lattice spacing $a_H$ of the lattice in which the spins in the Heisenberg model reside. Comparing (3.12) with (3.2) obtained from the AFHM, with $\rho_s = 0.227J$ and $C_\xi = 0.25a_H$, we find $a_0 = 2.69a_H$ and $T_0 = 0.392J$. Using the expression (3.11.b) we find $\hbar c_\sigma = 1.05Ja_H$. $c_\sigma$ is a parameter of the $\sigma$ model, not the physical spin-wave velocity.

The inverse correlation length versus $T$ as observed by neutron scattering experiments[15] done on the undoped $La_2CuO_4$, is compared with our results in Fig. 8. The solid curve is the exponential given by Eq. (3.2) which fits the results for $\xi(T)$ obtained from the nonlinear $\sigma$ model and AFHM. In the plot we used $a_H = 3.8Å$, the $Cu - Cu$ distance, and $J = 1480K$. This value of $J$ is close to the value reported by Raman scattering experiments[28]. Using this value of $J$ and the expression $c = 1.22\sqrt{2}Ja$ we obtain $c \simeq 0.84eV - Å$, a value higher than the lower bound of $0.6 eV Å$ reported by thermal neutron scattering studies[15] and closer to the more recent value of $0.85 eV Å$ inferred from high-energy inelastic neutron scattering[29].

## 4. Hole Dynamics

Following the mapping of the spin-$\frac{1}{2}$ AFHM to the quantum lattice-gas model of bosons, the problem of a hole in a quantum antiferromagnet (in the framework of the $t - J$ model) can be mapped to an impurity moving in a Bose system. Feynman and Cohen[30] have proposed a variational wave function for the motion of a $^3He$ impurity in liquid $^4He$. This wave function takes into account the background boson-boson short-range correlation due to the hard core interaction and impurity-boson backflow correlations. We have studied [18] the following variational wave function which includes spin-spin and spin-hole ("spin-backflow") correlations

$$|\Psi(\vec{k})> = \frac{1}{\sqrt{N}} \sum_{\vec{R}} e^{-i\vec{k}\cdot\vec{R}} \hat{G}_{\vec{k}} |\phi(\vec{R})>  \quad (4.1)$$

where $|\phi(\vec{R})>$ is the state (2.1.c) with a hole at $\vec{R}$. The correlation operator $\hat{G}_{\vec{k}}$ has the following form:

$$\hat{G}_{\vec{k}} = exp\Big(\sum_{\vec{\tau}} \big(ig_{\vec{k}}(\vec{\tau}) + h_{\vec{k}}(\vec{\tau})\big) S^z_{\vec{R}+\vec{\tau}} - \frac{1}{2}\sum_{i<j} u(r_{ij}) S^z_i S^z_j\Big). \quad (4.2)$$

The second term keeps the bosons from coming close to each other and from facing the hard core. The real function $h_{\vec{k}}(\vec{\tau})$ takes into account the hole-spin short-range correlations, while the term $ig_{\vec{k}}(\vec{\tau})$ represents the change in the phase of the wave function of the impurity in order to conserve local current in its motion.

Such backflow correlations in the case of a $^3He$ impurity in liquid $^4He$ are responsible for the large effective mass of the impurity. In the magnetic language the origin of the second term is the same as that in the pure AFHM and creates spin-fluctuations (the $|\phi>$ state is a Néel state in the $x$-direction therefore the $S^z_i S^z_j$ term creates pair-fluctuations). The first term couples the almost Néel state with the state where the hole hops to a neighboring site. In general $u_{ij}$ may depend on the distance from the hole. In practice we find no significant lowering of the hole energy by allowing for such a non-uniform $u_{ij}$. Therefore, we used the same $u_{ij}$ as that found in the VMC calculation at half-filling (AFHM) which is explained in section 2. Using the Metropolis algorithm we calculated[18] the expectation value of the $t-J$ Hamiltonian with the wave function (4.1-2) for one hole.

Fig. 9.a compares our variational results for the energy of a hole (we have subtracted the energy of the no-hole state) moving with $\vec{k} = (\frac{\pi}{2}, \frac{\pi}{2})$ for the two cases, $g_{\vec{k}} = 0$, and for the optimal value of $g_{\vec{k}}$, with exact diagonalization results on the $4\times 4$ lattice [31] for several values of $J/t$. The introduction of the spin-hole correlations ($g_{\vec{k}} \neq 0$) improves the energy significantly. The variational wave function is reasonably good for $J > t$; for lower values of $J/t$ we need to improve the wave function with the inclusion of three-body correlations. Similar results are found for the $10 \times 10$ lattice.

The single-hole energy $e(\vec{k})$ as a function of the momentum $\vec{k}$ is calculated for several values of $J/t$. A typical result is shown in Fig. 9.b calculated for $J/t = 1$ on a $10 \times 10$

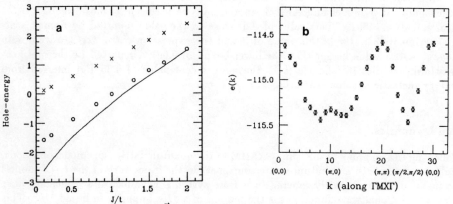

Figure 9. (a) Hole energy for $\vec{k} = (\frac{\pi}{2}, \frac{\pi}{2})$ as a function of $J/t$. The solid line is exact diagonalization results[31], while the crosses and circles are obtained with the variational wave function (4.1-2) with $g_{\vec{k}} = 0$ and $g_{\vec{k}} \neq 0$ respectively. (b) The hole-band $e(\vec{k})$.

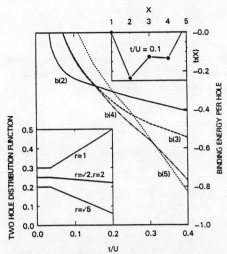

*Figure 10. The binding energies per hole $b(X)$ for a system of $X$ holes as a function of $t/U$. The lower-left inset gives the hole-hole distribution function for various distances. The top-right inset demonstrates the "magic" numbers for even number of holes.*

lattice. We have not subtracted the no-hole energy in this case. The curve has a minimum at $\vec{k} = (\frac{\pi}{2}, \frac{\pi}{2})$. The maximum value is at $\vec{k} = (0,0)$ or $\vec{k} = (\pi, \pi)$ and both values are the same within error bars. The bandwidth decreases with $J/t$. We also find, in agreement with other authors[31,32], that the effective mass of the hole in the direction towards $(0, \pi)$ is larger than that in the direction towards $(0, 0)$.

We have also calculated[18] the spin-hole-spin correlation function with one of the spins staying far away and the other close to the hole, using the variational wave function (4.1-2). In the static-hole case (t=0), we find that the staggered magnetization is enhanced in the neighborhood of the hole in accord with the results of Ref. 33. In the mobile-hole case (for $t/J \sim 1$), however, the staggered magnetization near the hole is significantly reduced from its bulk value in agreement with the spin-bag model of Schrieffer et al.[34].

Next, we shall discuss results obtained by exact diagonalization studies of the strong-coupling Hubbard Hamiltonian [19] on finite clusters of 10-sites. This Hamiltonian, in addition to the terms present in the $t - J$ model (1.1), involves a three-site hopping term. The full Hamiltonian is given in Ref. 19. First, it is also found that the minimum of the band in the AF phase is at $(3\pi/5, \pi/5)$, which is the closest wave vector to $(\pi/2, \pi/2)$ from the the 10 possible $\vec{k}$-vectors.

In Fig. 10 we plot the binding energy per hole obtained in Ref. 19 and defined as

$$b(X) = \frac{E(X) - XE(1)}{X} \tag{4.3}$$

where $E(X)$ is the total energy of the system containing $X$ holes, measured from the energy of the no-hole state. Notice that there is a range of $t/U$ ($0.04 < t/U < 0.16$) where two-hole pairing is favored against multi-hole clustering or phase separation. At larger values of $t/U$ multi-hole clustering is possible as can been seen in the figure (see also Ref. 19). The lower left inset shows the hole-hole distribution function and the top-right inset shows $B(X)$ versus $X$. The latter inset demonstrates that the peaks in $b(X)$ occur at even hole numbers.

## 5. Acknowledgements

This work was supported in part by the Florida State University Supercomputer Computations Research Institute which partially funded by the U.S. Department of Energy under Contract No. DE-FC05-85ER25000 and in part by the DARPA sponsored Florida Initiative in Advanced Microelectronics and Materials under Contract No. MDA972-88-J-1006.

## 6. References

1. J.E. Hirsch, Phys. Rev. Lett. **54**, 1317 (1985). K. Huang and E. Manousakis, Phys. Rev. **B36**, 8302 (1987).
2. N. D. Mermin and H. Wagner, Phys. Rev. Lett. **22**, 1133 (1966).
3. K. Kubo Phys. Rev. Lett. **61**, 110 (1988). K. Kubo and T. Kishi, Phys. Rev. Lett. **61**, 2585 (1988).
4. P. W. Anderson, Phys. Rev. **86**, 694 (1952). R. Kubo, Phys. Rev. **87**, 568 (1952). T. Oguchi, Phys. Rev. **117**, 117 (1960).
5. M. Parrinello and T. Arai, Phys. Rev. **B10**, 265 (1974). D. Huse, Phys. Rev. *B37*, 2380 (1988).
6. J. Oitmaa and D.D. Betts, Can. J. Phys. **56**, 897 (1978). J. D. Reger and A. P. Young, Phys. Rev **B37**, 5978 (1988). S. Miyashita, J. Phys. Soc. Jpn. **57**, 1934 (1988). T. Barnes, Phys. Rev. **B37**, 9405 (1988). Y. Okabe and M. Kicuchi, J. Phys. Soc. Jpn. **57**, 4351 (1988).
7. M. Gross, E. Sànchez-Velasco and E. Siggia, Phys. Rev. **B39**, 2484.
8. J. Carlson, Phys. Rev. **B40**, 846 (1989). N. Trivedi and D. Ceperley, *ibid*. **B40**, 2747 (1989).
9. E. Manousakis, Phys. Rev. **B40**, 4904 (1989).
10. Z. Liu and E. Manousakis, Phys. Rev. **B40**, 11437 (1989).
11. E. Manousakis and R. Salvador, Phys. Rev. Lett. **60**, 840 (1988).
12. E. Manousakis and R. Salvador, Phys. Rev. **B39**, 575 (1989).
13. E. Manousakis and R. Salvador, Phys. Rev. Lett. **62**, 1310 (1989); and Phys. Rev. **B40**, 2205 (1989).
14. F. D. M. Haldane, Phys. Lett. **93 A**, 464 (1983); Phys. Rev. Lett. bf 50, 1153 (1983). T. Dombre and N. Read, Phys. Rev. **B 38**, 7181 (1988); E. Fradkin and M. Stone, *ibid* **38**, 7215 (1988); L. B. Ioffe and A. I. Larkin, Int. J. Mod. Phys. **B 2**, 203 (1988); X.-G. Wen and A. Zee, Phys. Rev. Lett. **61**, 1025 (1988), F.D.M. Haldane, *ibid* **61**, 1029 (1988).
15. D. Vaknin, S.K.Sinha, D.E. Moncton, D.C. Johnston, J.M. Newsam, C.R. Safinya and H.E. King, Jr. Phys. Rev. Lett. **58**, 2802 (1987). G. Shirane, Y. Endoh, R.J.Birgeneau, M. A. Kastner, Y. Hidaka, M. Oda, M. Suzuki, and T. Murakami, Phys. Rev. Lett. **59**, 1613 (1987). Y. Endoh, et al., Phys. Rev., **B37**, 7443 (1988). K. Yamada, K. Kakurai, Y. Endoh, T. R. Thurston, M. A. Kastner, R. J. Birgeneau. G. Shirane, Y. Hidaka and T. Murakami, Preprint.
16. S. Chakravarty, B.I. Halperin, and D. Nelson, Phys. Rev. Lett. **60**, 1057 (1988).
17. S. Chakravarty, B.I. Halperin, and D. Nelson, Phys. Rev. **B39**, 2344 (1989).
18. M. Boninsegni and E. Manousakis, to be published. See also *Quantum Simulations of Condensed Matter Phenomena*, Los Alamos, NM, August 8-11, 1989, (World Scientific).
19. E. Kaxiras and E. Manousakis, Phys. Rev. **B38**, 866 (1988) and Phys. Rev. **B37**, 656 (1988).
20. L. Hulthen, Ark. Mat. Astr. Fys., **26A**, No. 1 (1938). P. W. Kastelijn, Physica **18**, 104 (1952). W. Marshall, Proc. R. Soc. London, Ser. A **232**, 48 (1955). R. Bartkowski, Phys. Rev. **B5**, 4536 (1972).

21. D. A. Huse and V. Elser, Phys. Rev. Lett. **60**, 2531 (1988). P. Horsch and W. von der Linden: Z. Phys. **B72**, 181 (1988).
22. G. V. Chester and L. Reatto, Phys. Lett. **22**, 276 (1966).
23. H. Neuberger, T. Ziman, Phys. Rev. **B39**, 2608 (1989). M. Gross, E. Sánchez-Velasco and E. Siggia, Phys. Rev. **B39**, 2484 (1989).
24. R. R. P. Singh, P. A. Fleury, K. B. Lyons, P. E. Sulewski, Phys. Rev. Lett. **62**, 2736 (1989).
25. M. Gross, E. Sánchez-Velasco and E. Siggia, Preprint.
26. D. P. Arovas and A. Auerbach, Phys. Rev. **B38**, 316 (1988). M. Takahashi, Phys. Rev. **B40**, 2494 (1989),
27. E. Brèzin, J. Physique, **43**, 15 (1982).
28. K. B. Lyons, P.A. Fleury, J.P.Remeika and T.J. Nergan, Phys. Rev. **B37**, 2353 (1988).
29. G. Aeppli, et al., Phys. Rev. Lett. **62**, 2052 (1989).
30. R. P. Feynman and M. Cohen, Phys. Rev. **102**, 1189 (1957).
31. E.Dagotto, A.Moreo, R.Joynt, S.Bacci, and E. Gagliano, NSF-ITP-89-74, May, 1989. C. -X. Chen and H.-B. Schüttler, Univ. of Georgia, Preprint, 1989.
32. B.Shraiman and E.Siggia, Phys. Rev. Lett. **60**, 740 (1988). C.Gros and M.D.Johnson, to be published. C.L.Kane, P.A. Lee, and N.Read, Phys. Rev. **B39**, 6880 (1989). S. Schmitt-Rink, C. M. Varma, and A. E. Ruckenstein, Phys. Rev. Lett. **60**, 2793 (1988).
33. N. Bulut, D. Hone, E. Loh and D. Scalapino, Phys. Rev. Lett. **62**, 2192 (1989).
34. J. R. Schrieffer, X. -G. Wen and S. -C. Zhang, Phys. Rev. Lett. **60**, 944 (1988).

# Binding of Holes in the Hubbard Model

A. Moreo

Department of Physics, University of California, Santa Barbara, CA 93106, USA

**Abstract.** We present a new numerical method for the study of binding energies of particles in fermionic systems. Working with an *imaginary* chemical potential we can obtain results in the *canonical* ensemble by simple modifications of standard numerical techniques. We applied the technique to the two dimensional (2D) Hubbard model observing binding of holes at half-filling on lattices of $4 \times 4$ sites. For $U/t = 4$ we estimate that the binding energy is $\Delta = -0.10 \pm 0.02$ [1].

The two dimensional Hubbard model is being studied extensively in relation to the high temperature superconductors [2], and it appears to describe some of the features of these new materials. However, it remains an open question whether this model actually has a superconducting phase. Analytic studies of the Hubbard model are difficult in the intermediate and strong coupling regimes where perturbative and mean field techniques are questionable. Numerical simulations are difficult at low temperatures for the relevant band fillings because the integrands of the Feynman path integrals are not positive definite. In this talk we use a new numerical technique to show that two holes in a half-filled sea do bind for a $4 \times 4$ lattice. This result provides evidence that an effective attractive interaction appears in the $U > 0$ Hubbard model. Our results do not constitute a proof that a superconducting phase will appear at finite hole doping, but the existence of a net attractive force between holes makes that scenario more plausible. As the criterion for binding we use the quantity,

$$\Delta = (E_2 - E_0) - 2(E_1 - E_0) = (E_2 - E_1) - (E_1 - E_0), \tag{1}$$

where $E_n$ denotes the energy of the ground state with $n$ holes (i.e. $n$ less fermions than a half-filled band). If there is a bound state of two holes then $\Delta < 0$. A calculation of $\Delta$ should be done in the canonical (C) ensemble where the number of fermions is fixed, rather than in the grand canonical (GC) ensemble. The Lanczos method can be used for this purpose, and it has produced interesting results for the $t-J$ and Heisenberg models[3], and very recently for the Hubbard model [4]. However, it can not be used for large lattices. Stochastic methods have recently been proposed for calculating ground state properties with fixed numbers of electrons [5,6]. However, they suffer from the sign problem mentioned above. In addition, they provide a direct calculation of the (extensive) energies $E_n$. This is a problem in evaluating $\Delta$, which is an intensive quantity, since one must take differences of extensive quantities, which requires high precision measurements. Our alternative numerical approach is based on a simple modification of the standard Quantum Monte Carlo method used in calculations of the GC-ensemble [6,7]. Although below we focus our attention specifically on the Hubbard model, the derivation is valid for a wide variety of Hamiltonians.

Consider a model defined by a Hamiltonian $\hat{H}$. We denote the partition function for the GC-ensemble with chemical potential $\mu$ by $Z_{GC}(\mu)$ and the partition function for the C-ensemble with $n$ electrons above half-filling by $Z_C(n)$. They are related by

$$Z_{GC}(\mu) = tr[e^{-\beta(\hat{H}-\mu\hat{N})}] = e^{\beta\mu N} \sum_{n=-N}^{n=N} e^{\beta\mu n} Z_C(n), \quad (2)$$

where $\hat{N}$ is the number operator and $N$ the number of spatial lattice points. For the Hubbard model the eigenvalues of $\hat{N}$ range from 0 to $2N$. Our approach rests on the continuation of Eq. 2 to imaginary chemical potential, $\mu \to i\lambda$. Since the canonical partition functions are independent of $\mu$,

$$Z_C(n) = \frac{1}{2\pi\beta} \int_0^{2\pi/\beta} d\lambda \, e^{-i\beta\lambda(n+N)} Z_{GC}(\mu = i\lambda). \quad (3)$$

The advantage of Eq. 3 is that for the Hubbard model, and a variety of other models, $\exp(-i\beta\lambda N)Z_{GC}(\mu = i\lambda)$ can be written as a path integral with a *positive* weight. All the problems associated with fluctuating signs are contained in the explicit phase factor $\exp(-i\beta\lambda n)$ rather than in an involved determinant phase as in previous formulations.

We now specialize to the Hubbard model defined by the Hamiltonian,

$$\hat{H}(t,U) = -t \sum_{\substack{<i,j> \\ \sigma}} (c_{i,\sigma}^\dagger c_{j,\sigma} + h.c.) + U \sum_i (n_{i,\uparrow} - \tfrac{1}{2})(n_{i,\downarrow} - \tfrac{1}{2}), \quad (4)$$

where the notation is standard. To perform a numerical calculation we integrate out the fermion degrees of freedom. To this end we use a discrete Hubbard-Stratonovich transformation [8], introducing an imaginary-time dependent Ising variable $\{s\}$ at each lattice site. The spin up and down electrons couple to the $\{s\}$ fields in proportion to $n_{i,\sigma}$ with *opposite* signs. Integrating out the fermion degrees of freedom we write the GC-ensemble partition function as a sum over configurations, $\{s\}$ [6,7]

$$Z_{GC}(\mu) = \sum_{\{s\}} det M(\{s\},\mu)_\uparrow det M(\{s\},\mu)_\downarrow, \quad (5)$$

where $M_\uparrow$ and $M_\downarrow$ are the fermion matrices for spin up and down electrons propagating through the configuration $\{s\}$ in a chemical potential $\mu$. We now perform a particle–hole transformation on the spin down sector given by $c_{i,\downarrow} \to c_{i,\downarrow}^\dagger(-1)^{i_x+i_y}$, $i = (i_x, i_y)$. Under this transformation the coupling of the spin up and down electrons to the $\{s\}$ field become identical and $\mu(n_{i,\uparrow} + n_{i,\downarrow}) \to \mu(n_{i,\uparrow} - n_{i,\downarrow} + 1)$. Eq. 5 can therefore be rewritten as

$$Z_{GC}(\mu) = e^{-\beta\mu N} \sum_{\{s\}} \rho det M(\mu)_\uparrow det M(-\mu)_\uparrow, \quad (6)$$

where $\rho$ is a positive function of the $\{s\}$ variables [8]. Now each term in the sum Eq. 6 is positive for $\mu = i\lambda$, since each determinant is the complex conjugate of the other.

We are primarily interested in calculating the binding energy, $\Delta$, and we wish to directly obtain the energy differences on the right hand side of Eq. 1 to avoid cancellations in subtracting two large quantities. This can be done using the asymptotic result

$$\frac{Z_C(n)}{Z_C(0)} \longrightarrow d_{n,0} e^{-\beta(E_n - E_0)}. \tag{7}$$

where $d_{n,0}$ is the ratio of degeneracies of the ground states with $n$ and $0$ holes. A straightforward way to proceed would be to generate a sequence of field configurations, $\{s\}$ and complex chemical potential values $\lambda$, with a probability distribution, $P(\{s\}, \lambda)$, proportional to $\rho |\det M(\{s\}, i\lambda)_\uparrow|^2$. $Z_C(n)/Z_C(0)$ is given by the expectation value of $\exp(-i\beta\lambda n)$ in this distribution (particle–hole symmetry implies that $Z_C(n) = Z_C(-n)$, so it is only necessary to measure $\cos(\beta\lambda n)$). However, because a gap exists in the single electron density of states, the fermion determinant is nearly $\lambda$-independent at low temperatures. Then, we can simply generate the field configurations for $\lambda = 0$, just as one does in GC-ensemble calculations at $\mu = 0$. We then have

$$\frac{Z_C(n)}{Z_{GC}(0)} = \sum_{\{s\}} P(0) \frac{1}{2\pi\beta} \int_0^{2\pi/\beta} d\lambda \, e^{-i\beta\lambda n} \frac{|\det M(i\lambda)_\uparrow|^2}{|\det M(0)_\uparrow|^2}. \tag{8}$$

We do not need to explicitly evaluate the integral over $\lambda$ in Eq. 8. The absolute square of the fermion determinant is simply the partition function for electrons in the presence of the $\{s\}$ field, so we can write in analogy with Eq. 2

$$\frac{|\det M(i\lambda)_\uparrow|^2}{|\det M(0)_\uparrow|^2} = c_0(\{s\}) + 2\sum_{n=1}^{N} \cos(\beta\lambda n) \, c_n(\{s\}), \tag{9}$$

where we used the fact that the left hand side of Eq. 9 is even under $\lambda \to -\lambda$. Clearly $\frac{Z_C(n)}{Z_{GC}(0)} = \langle c_n \rangle$ where the average is over an ensemble of $\{s\}$ configurations generated by the probability $P(\{s\}, 0)$. To obtain the gaps $\Delta_{nm} = E_n - E_m$ we then need,

$$\frac{Z_C(n)}{Z_C(m)} = \frac{\langle c_n \rangle}{\langle c_m \rangle}. \tag{10}$$

Since each $Z_C(m) \propto \exp(-\beta E_m)$ and $Z_{GC}(0) \propto \exp(-\beta E_0)$ for large $\beta$, Eq. 10 involves intensive quantities, and no large cancellations occur. To obtain the $c_n$ for a given field configuration, we evaluate the fermion determinant for an arbitrary set of $\lambda_i$ and invert [9] Eq. 9. Numerical calculations cannot be carried out at arbitrarily low temperatures since the quantities $Z_C(n)/Z_{GC}(0)$ fall off exponentially with $\beta$ in part because the $c_n$ become small configuration by configuration, and also because of cancellations between configurations ($c_{n>0}$ can be negative for some configurations of $\{s\}$) The latter effect introduces noise that eventually makes the calculations impractical. For example, for $4 \times 4$ lattices we are restricted to $\beta \leq 6$. What we have done is to perform calculations for that range of $\beta$, and then perform fits to obtain the gaps.

We checked our technique on a $2 \times 2$ lattice where exact results can be obtained. We used a time step $\Delta \tau = 0.075$ and between $2000 - 5000$ iterations after thermalization. Numerically evaluating $-log(\frac{\langle c_2 \rangle}{\langle c_1 \rangle})$ vs. $\beta$ the agreement with the exact results at each temperature was excellent. The asymptotic regime Eq. 7 is easily reached and from there we obtained $\Delta_{21} = 1.16 \pm 0.01$ (in units of $t$) while the exact result is 1.1640. For the case $\frac{\langle c_1 \rangle}{\langle c_0 \rangle}$ special care must be taken. Although the agreement with the exact values at each temperature is also excellent, the asymptotic regime is not easily reachable and important deviations from it can be seen even at relatively low temperatures. The reason is that in the half-filled subspace there is an energy level very close to the ground state (that can be mapped into the spin-wave of strong coupling) perturbing the asymptotic result Eq. 7. The remedy to this problem is to fit the data including this 'spin-wave' level, which is three-fold degenerate, with its energy ($\Delta_{SW}$) as a free parameter, as

$$\frac{Z_1}{Z_0} = \frac{\langle c_1 \rangle}{\langle c_0 \rangle} \longrightarrow \frac{d_{1,0} e^{-\beta \Delta_{10}}}{1 + 3 e^{-\beta \Delta_{SW}}}. \tag{11}$$

After this improvement the data can be fit very well (Fig. 1) using $d_{1,0} = 4$. The prediction for the gap is $\Delta_{10} = 1.24 \pm 0.01$ while the exact result is 1.2383. The numerical spin-wave gap is $\Delta_{SW} = 0.29 \pm 0.01$ (exact result $\Delta_{SW} = 0.2815$).

Now we present results for the $4 \times 4$ lattice. For the gap $\Delta_{10}$ we follow the same technique used for the $2 \times 2$ lattice. In Fig. 2, we show $-Tln(\frac{<c_1>}{12<c_0>})$ vs. $T$ at $U/t = 4$. Our numerical results are $\Delta_{10} = 0.98 \pm 0.02$ and $\Delta_{SW} = 0.47 \pm 0.03$. In Fig. 3 we show $-log(\frac{\langle c_2 \rangle}{\langle c_1 \rangle})vs.\beta$ at $U/t = 4$. A good straight-line behavior is observed obtaining a gap $\Delta_{21} = 0.88 \pm 0.02$. The optimal value of $d_{2,1}$ is $\approx 1.5$

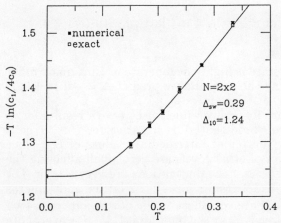

Fig. 1: $-Tlog(\frac{<c_1>}{4<c_0>})$ vs. $T$ for the $2 \times 2$ lattice with $U/t = 4$. The solid line is our extrapolation using Eq. 11 with $d_{1,0} = 4$.

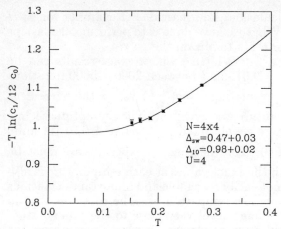

Fig. 2: $-T log(\frac{<c_1>}{12<c_0>})$ as a function of $T$ for the $4 \times 4$ lattice with $U/t = 4$. The solid lines are our extrapolations using Eq. 11 with $d_{1,0} = 12$.

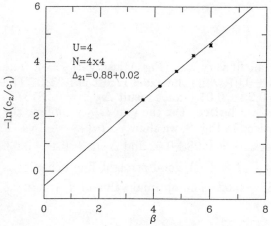

Fig. 3: $-log(\frac{\langle c_2 \rangle}{\langle c_1 \rangle})$ as a function of $\beta$ for the $4 \times 4$ lattice with $U/t = 4$.

implying that the two hole ground state is highly degenerate or that many other levels are very close to it. Combining these results we predict $\Delta = -0.10 \pm 0.02$ at $U/t = 4$.

We have repeated this procedure for many values of $U/t$ with results for $\Delta$ shown in Fig. 4. We can safely conclude that $\Delta$ is negative for $U \leq 7$ on a $4 \times 4$ lattice. Our technique works better at intermediate values of $U/t$ since for $U/t \ll 1$, $\Delta_{SW}$ and the energies of other levels are very small affecting the extrapolation to $T = 0$ while for $U/t \gg 1$ the fluctuations are very strong. For comparison, in Fig. 4 we also show exact results for a $2 \times 2$ lattice and Lanczos results for a $\sqrt{8} \times \sqrt{8}$ lattice [10]. Both are very close to the new results on a $4 \times 4$ lattice suggesting that we may be already near the bulk limit [11]. However, note that at least for $U/t = 4$, $\Delta$ seems to be smaller for the 16 sites lattice than for the 8 site lattice. Then our result $|\Delta| = 0.10 \pm 0.02$ is perhaps an upper bound

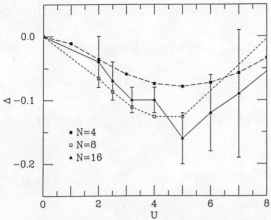

Fig. 4: $\Delta$ as a function of $U/T$ for the $2 \times 2$ (long-dashed line), $\sqrt{8} \times \sqrt{8}$ (short-dashed line) and $4 \times 4$ (continuous line) lattices.

for the bound state energy. For large $U/t$, $\Delta$ is positive, meaning that the holes prefer to be separate as much as possible in the small lattice we studied (in the bulk limit $\Delta$ can not be positive). That roughly happens for values of $U/t \geq 8$ where the holes may prefer to form independent ferromagnetic polarons. So the interesting regime in this model appears to be $U/t \approx 4 - 5$ rather than larger values.

After completing Fig. 4 we received a paper by Parola et al.[4] in which Lanczos and stochastic results for the $4 \times 4$ lattice are presented for 2 and 0 holes at $U/t = 4$. The Lanczos result for this gap is $\Delta_{20} = 1.8784$ while our prediction is $\Delta_{20} = \Delta_{21} + \Delta_{10} = 1.87 \pm 0.02$ in excellent agreement within statistical errors. The stochastic result of Parola et al. is [12] $\Delta_{20} = 1.76 \pm 0.08$. which was obtained by directly measuring the extensive energies rather than energy differences.

Acknowledgments

This work was supported by NSF grants PHY86-14185 and PHY82-17853, supplemented by funds from NASA, and by DOE grant FG03-85-ER45197. The computer simulations were performed on the Cray-2 of NCSA and at SDSC.

References

1. E. Dagotto, A. Moreo, R. Sugar, and D. Toussaint, Phys. Rev. B **41**, 811 (1990).

2. J.G. Bednorz and K.A. Müller, Z. Phys. **B64**, 188 (1986); P.W. Anderson, Science **235**, 1196 (1987).

3. E. Dagotto and A. Moreo, Phys. Rev. Lett. **63**, 2148 (1989); E. Dagotto et al., NSF–ITP–89–74 and references therein.

4. A. Parola, S. Sorella, S. Baroni, R. Car, M. Parinello, and E. Tosatti, Trieste preprint.

5. S. Sorella, S. Baroni, R. Car, and M. Parrinello, *Europhys. Lett.* **8**, 663 (1989).

6. S.R. White, D.J. Scalapino, R.L. Sugar, E.Y. Loh, J.E. Gubernatis, and R.T. Scalettar, *Phys. Rev. B* **40**, 506 (1989).

7. R. Blankenbecler, D.J. Scalapino, and R.L. Sugar, *Phys. Rev. D* **24**, 2278 (1981).

8. J.E. Hirsch, *Phys. Rev. B* **31**, 4403 (1985).

9. The $c_n$ fall off rapidly with $n$ at low temperatures, so the sum can be truncated, and the fermion determinant need only be evaluated for a limited number of $\lambda_i$ (on the $4 \times 4$ lattice we needed only five terms).

10. J. Riera and A.P. Young, *Phys. Rev. B* **39**, 9697 (1989).

11. We do not expect phase separation in this model based on the results for an 8 site lattice of Ref. 10.

12. We follow the convention that the error bars of a sum of energies is the biggest of the two error bars.

# The Average Spectrum Method for the Analytic Continuation of Imaginary-Time Data

*S.R. White*

Department of Physics, University of California, Irvine, CA 92717, USA

In this paper we present the average spectrum method, a new method for obtaining real-frequency information from imaginary-time quantum Monte Carlo data. This technique does not require the adjustable parameters, smoothness constraints, or model forms of some previous techniques, yet produces smooth, consistent spectra from noisy data. Various tests of the method on mock data are presented, as well as realistic applications to the two-dimensional Hubbard model.

## 1. Introduction

Many of the most interesting observables of a many-electron system—densities of states, frequency-dependent susceptibilities, and the conductivity, for example—are *dynamical* quantities, *i.e.*, they are functions of real frequency. Unfortunately, it is usually much easier to calculate static or imaginary-time correlation functions than to calculate functions of real time or frequency. This is particularly true for finite-temperature fermion Monte Carlo, which has had considerable success in studies of strongly interacting electron models such as the Hubbard model. With this technique, imaginary-time dependent quantities can be evaluated directly, but dynamic quantities can only be obtained by analytically continuing the imaginary-time data. The statistical errors inherent in the Monte Carlo data make this analytic continuation especially difficult. Nevertheless, the importance of the information that can be obtained even from fairly low resolution dynamical spectra, and the lack of alternative ways of obtaining these spectra, have prompted a flurry of recent work in developing new techniques for performing this analytic continuation. In this paper we describe a new continuation technique and test the method on both simple test cases and real Monte Carlo data.

Until recently, only a few attempts had been made to perform this analytic continuation. In one of the first approaches, Padé approximants were used to analytically continue quantities calculated from imaginary time Monte Carlo simulations for models with a single degree of freedom and for magnetic impurities.[1,2] For more complicated models, however, the Padé technique has insufficient resolution and stability. Schüttler and Scalapino developed a more stable technique, based on a least squares fitting procedure, and applied it to a system of spinless interacting fermions in one dimension.[3] Unfortunately, this technique gave largely qualitative, low resolution results. Within the last year, however, several advances have occurred in this area. White, Scalapino, Sugar, and Bickers[4] (hereafter denoted by WSSB) developed a procedure based on least squares which found the *smoothest* positive-definite real-frequency spectrum which was *consistent* with the imaginary-time data. This procedure was capable of high-resolution results, provided the statistical errors in the data were sufficiently small. This procedure was used to obtain densities of states and spectral weight functions for the two-dimensional (2D) Hubbard model. Independently Jarrel and Biham[5] developed a different method which also incorporated a bias towards smooth results. They applied this method to the Anderson impurity model. The development of these algorithms prompted Silver, Sivia, and Gubernatis[6] to apply a method often used in image reconstruction, the maximum entropy method, to the problem. An important advantage of this approach is that it is based on probability theory, in particular Bayes' Theorem. This theorem provides a very natural foundation for the problem of analytically continuing imaginary-time Monte Carlo data, and it turns out that procedures based on least-squares, such as those of Schüttler and Scalapino, and WSSB can be formulated in its framework.

The average spectrum method, which we introduce here, is also based on Bayes' Theorem. We believe this method has significant advantages over the techniques discussed above, the primary advantage being its conceptual simplicity. It gives stable, smooth results without the need to introduce any smoothness constraints, model forms, or any adjustable parameters. A brief description of the average spectrum method was first given in a short work by us[7], and was used to calculate dynamical pair susceptibilities of the 2D Hubbard model. In this paper we will discuss this method in more detail, including computational details. In Section 2 we discuss the foundation of the method in terms of probability theory, and in Section 3 we give some computational details. Section 4 gives results of tests of the method on Monte Carlo imaginary-time data as well as mock data sets, and recent results on pairing susceptibilities.

## 2. Bayes' Theorem and Analytic Continuation

We will consider the analytic continuation of the single-particle imaginary-time Green's function

$$G(p, \tau) \equiv -\langle T_\tau [c_p(\tau) c_p^\dagger(0)] \rangle \tag{1}$$

where $p$ runs over the set of discrete momenta allowed on a finite lattice. Recent advances in finite-temperature quantum Monte Carlo techniques[8,9] allow one to calculate $G(p,\tau)$ at low temperatures for models such as the half-filled Hubbard model. The imaginary-frequency Green's function for Matsubara frequency $i\omega_n$ is given by

$$G(p, i\omega_n) = \int_0^\beta d\tau G(p, \tau) e^{i\omega_n \tau} \tag{2}$$

with $\omega_n = (2n+1)\pi T$, $n = 0, \pm 1, \pm 2, \ldots$. The real frequency, retarded Green's function $G_{\text{ret}}(p,\omega)$ is formally obtained from $G(p, i\omega_n)$ by the continuation $i\omega_n \to \omega + i\delta$. The spectral weight function is defined as

$$A(p, \omega) = -\frac{1}{\pi} \text{Im} \, G_{\text{ret}}(p, \omega), \tag{3}$$

and is positive definite. The single-particle density of states is $N(\omega) = 1/N \sum_p A(p, \omega)$. Inserting complete sets of many-particle eigenstates into (1)-(3), it is straightforward to show that for $0 \leq \tau \leq \beta$

$$G(p,\tau) = -\int_{-\infty}^{\infty} d\omega \frac{A(p,\omega)e^{-\tau\omega}}{1+e^{-\beta\omega}} \equiv -\mathcal{L}\{A\}. \tag{4}$$

From the quantum Monte Carlo calculation one typically obtains $G(p,\tau)$ at a discrete set of points $\tau_\ell = \ell\Delta\tau$, $\ell = 0,\ldots,L$, where $L\Delta\tau = \beta$. One also can obtain an estimate of the statistical error $\sigma(p,\tau_\ell)$ for each data point; for now we will assume the errors are independent for different values of $\tau_\ell$, and postpone discussion of correlations in the errors until Section 3.

If the exact value of $G(p,\tau)$ is $G^{\text{ex}}$, and one assumes the statistical error has a Gaussian distribution of width $\sigma$, then the probability of obtaining the value $G^{\text{data}}$ from the Monte Carlo is proportional to

$$\Pr[G^{\text{data}}] \propto \exp(-\frac{1}{2}(G^{\text{data}} - G^{\text{ex}})^2/\sigma^2). \tag{5}$$

Similarly, if the exact value of $A(p,\omega)$ is $A^{\text{ex}}(p,\omega)$, the probability of obtaining the complete set of data $G^{\text{data}}$ given $A^{\text{ex}}(p,\omega)$ is

$$\Pr[G^{\text{data}}|A^{\text{ex}}] \propto \exp(-\frac{1}{2}\sum_\ell \epsilon_\ell^2/\sigma_\ell^2) \tag{6}$$

where
$$\epsilon_\ell = G^{\text{data}}(\tau_\ell) + \mathcal{L}\{A^{\text{ex}}\}(\tau_\ell). \tag{7}$$

Bayes' Theorem tells us how to obtain the probability that the exact spectral weight is $A(p,\omega)$ given the data $G^{\text{data}}$ from the probability $\Pr[G^{\text{data}}|A]$. It states that

$$\Pr[A|G^{\text{data}}] \propto \Pr[G^{\text{data}}|A]\Pr[A]. \tag{8}$$

Here $\Pr[A]$, termed the prior probability, contains what information we know about $A$ in the absence of the data; e.g., that $A$ is positive definite.

Both the WSSB method and the maximum entropy method are *maximum likelihood methods*, in that they calculate the $A$ which maximizes $\Pr[A|G^{\text{data}}]$. In order to obtain results which show only statistically significant features in a maximum likelihood method, it is necessary to incorporate biases in $\Pr[A]$ which favor smooth, featureless results, so that the data must force the significant features to appear. The methods above differ in their choice of $\Pr[A]$: the WSSB method asserts that *smooth* spectra are more likely than rapidly varying spectra; the maximum entropy method asserts that spectra which are closer to a model form (typically $\omega = $ const.) are more likely. In both cases a parameter in $\Pr[A]$ is adjusted to allow only statistically significant features to appear in the result. In the WSSB method this parameter is adjusted in a somewhat *ad hoc* basis, while in the maximum entropy method the parameter is set using probability arguments.

In the average spectrum method, we avoid the need to introduce biases in $\Pr[A]$ and avoid any adjustable parameters by abandoning the maximum likelihood method and instead *average* over $\Pr[A|G^{\text{data}}]$. We calculate the average spectrum via

$$\langle A(p,\omega)\rangle = \int \mathcal{D}A(p,\omega)\Pr[A|G^{\text{data}}]A(p,\omega) \tag{9}$$

with
$$\Pr[A] = \begin{cases} \text{const.,} & \text{if } A(p,\omega) \geq 0 \text{ for all } \omega; \\ 0, & \text{otherwise.} \end{cases} \quad (10)$$

The integral here is a path integral over all positive definite spectra. Each path in the integral is a possible spectrum weighted by the probability of that spectrum, given the data. The averaging over spectra automatically smears out statistically insignificant features. Because of this we need only include in $\Pr[A]$ what we really know: that the result is positive definite and in some cases that it is an even function. The average spectrum method makes fewer assumptions and is conceptually simpler than any of the maximum likelihood methods; the only disadvantage is that the path integral may be somewhat more time-consuming to compute. In the next section we discuss numerical techniques for performing the path integral which make the computation time for the analytic continuation small compared to typical calculation times needed for obtaining the data $G^{\text{data}}$.

## 3. Numerical Techniques

We represent the spectral density by a closely spaced set of delta functions with weights $A_i$
$$\bar{A}(p,\omega) = \sum_i A_i \delta(\omega - \omega_i). \quad (11)$$

where the $\omega_i$ are spaced uniformly: $\omega_i = -\Omega, -\Omega + \Delta\omega, \ldots, \Omega$. Then (4) becomes
$$\bar{G}(p,\tau) = -\sum_i A_i e^{-\tau\omega_i}/(1 + e^{-\beta\omega_i}), \quad (12)$$

The path integral in (9) then becomes an integral over each $A_i$
$$\int \mathcal{D}A(p,\omega) \approx \int_0^\infty dA_1 \int_0^\infty dA_2 \ldots \quad (13)$$

The probability function in the path integral becomes
$$\Pr[A|G^{\text{data}}] \propto \exp[-\frac{1}{2}(A \cdot M \cdot A - 2A \cdot g)] \equiv \exp[-\frac{1}{2}S] \quad (14)$$

with
$$M_{ij} = \sum_\ell \frac{1}{\sigma_\ell^2} f_{i\ell} f_{j\ell}, \quad (15)$$

$$g_i = \sum_\ell \frac{1}{\sigma_\ell^2} f_{i\ell} G^{\text{data}}(p, \tau_\ell), \quad (16)$$

$$f_{i,\ell} = e^{-\tau_\ell \omega_i}/(1 + e^{-\beta\omega_i}).$$

Because of the positivity constraint in (10), in order to do the path integral one must use Monte Carlo techniques. This Monte Carlo evaluation of the path integral is completely separate from the previous simulation of $G^{\text{data}}$, and generally takes much less computer time. The Monte Carlo proceeds as follows: first, one proposes a change in $A$: $A' = A + \alpha \Delta A$. In the usual practice $\Delta A$ would be nonzero at only a single frequency $\omega_i$; however, this type of move is highly inefficient because the total area of the spectrum is usually tightly constrained by the data. We choose $\Delta A$ to be nonzero on either (a) adjacent pairs of pixels; (b) nonadjacent pairs of pixels chosen at random; or (c) adjacent triplets of pixels. In cases (a) and (b) $\Delta A$ is $+1$ on one site and $-1$ on the other. For case (c) the relative weights of $\Delta A$ on each of the nonzero pixels is chosen by diagonalizing the $3 \times 3$ submatrix of $M$ corresponding to those pixels and setting $\Delta A$ to be the minimum eigenvector. (The diagonalization is done once for each triplet at the beginning of the calculation.) One then sets the coefficient $\alpha$ by the heat bath technique. Choosing the minimum eigenvector for $\Delta A$ in case (c) ensures that $\alpha$ will not be constrained to be nearly zero.

Usually correlations are quite pronounced in the statistical fluctuations between data points for different values of $\tau_\ell$. We have found that it is important to statistically estimate the correlations, rather than assume independence of the statistical errors, if one is to obtain the most reliable results. In that case one replaces the variance $\sigma_\ell^2$ by a covariance matrix $C_{\ell\ell'}$. We will discuss ways of estimating the covariance matrix in a subsequent paper.

## 4. Applications

Figure 1 shows results from an application of the average spectrum method to an artificial data set constructed from an exact set of $A_i$, forming $G^{\text{data}}$ using (12), and adding random statistical errors to the data. The exact $A$ consists of four delta-function peaks, all of equal height, at $\pm 1, \pm 2$. The root-mean-square error $\sigma$ of the noise added to the data is shown for each curve. For accurate resolution of each peak, an error of $10^{-5}$ is necessary. However, for larger values $\sigma$ one still obtains important features of the spectrum. The resolution is best near $\omega = 0$. Figure 2 shows a particular spectrum chosen from the ensemble for $\sigma = 10^{-4}$ at random. Whereas the average spectrum for $\sigma = 10^{-4}$ resolves each of the peaks, the "snapshot" spectrum shows numerous extraneous peaks which are not statistically significant.

Figures 3 and 4 show the density of states $N(\omega)$ of the $8 \times 8$ single band Hubbard model for $U = 4$ at half-filling and away from half-filling. (The hopping parameter $t$ has been set to unity.) The results at half-filling clearly show the insulating gap in $N(\omega)$ at $\omega = 0$; they also indicate quasiparticle peaks at the edges of the gap. In contrast, away from half-filling there is no gap; the system is a conductor. Simulations at lower temperature are possible at half-filling, where there is no fermion sign problem.[10] The slight roughness in the curve at half-filling near $\omega = 3$ is due to statistical errors in the evaluation of the path integral (9); longer running would smooth out the results.

Recently we have applied the average spectrum method to the calculation of dynamical pair susceptibilities of the 2D Hubbard model.[7] These susceptibilities are obtained by analytically continuing the imaginary-time correlation function

$$\chi(\tau) = \langle \Delta(\tau) \Delta^\dagger(0) \rangle \tag{17}$$

where

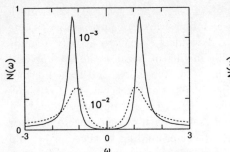 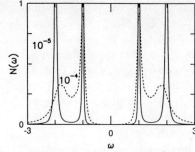

Fig. 1. Average spectrum results for statistical noise levels of $10^{-2}$, $10^{-3}$, $10^{-4}$, and $10^{-5}$. Here $\beta = 10$ and $\Delta\tau = 0.125$.

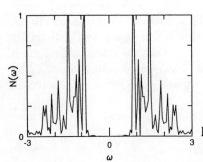 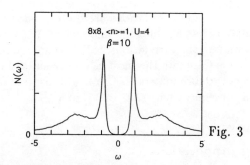

Fig. 2. A particular spectrum taken randomly from the probability distribution of spectra for a noise level of $10^{-4}$ in the test case of Fig. 1.

Fig. 3. Density of states for the 2D Hubbard model at half-filling, with $U = 4$ and $\beta = 10$.

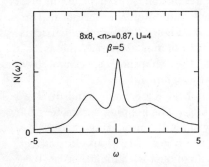

Fig. 4. Density of states for the 2D Hubbard model somewhat away from half-filling, with $U = 4$ and $\beta = 5$.

$$\Delta_f^\dagger = \frac{1}{\sqrt{N}} \sum_{i,j} f(i-j) c_{i\uparrow}^\dagger c_{j\downarrow}^\dagger \qquad (18)$$

where $f(i-j)$ is the pair wavefunction: $f(i-j) = \delta_{ij}$ for s-wave, $f(i-j) = 1/2[\delta_{i+\hat{x},j} + \delta_{i-\hat{x},j} \pm (\delta_{i+\hat{y},j} + \delta_{i-\hat{y},j})]$ for extended s-wave ($s^*$, +), and d-wave (−). The power spectrum $\bar{\chi}(\omega)$ is related $\chi(\tau)$ via

$$\mathcal{L}\{\bar{\chi}(\omega)\} = \chi(\tau) \qquad (19)$$

and is related to the retarded dynamic susceptibility by

$$\bar{\chi}(\omega) = \pi^{-1}\coth(\beta\omega/2)\text{Im}\chi^{\text{ret}}(\omega). \quad (20)$$

By expanding in complete sets of many-particle eigenstates $|s\rangle$ with energies $E_s$, we find

$$\bar{\chi}(\omega) = \sum_{s,s'}\frac{e^{-\beta E_s}}{Z}[|\langle s'|\Delta^\dagger|s\rangle|^2\delta(\omega-(E_{s'}-E_s))+|\langle s'|\Delta|s\rangle|^2\delta(\omega+(E_{s'}-E_s))] \quad (21)$$

where $Z$ is the partition function.

We see from (21) that at low temperatures $\bar{\chi}(\omega > 0)$ gives information on the state $\Delta_f^\dagger|N\rangle$, where $|N\rangle$ is the $N$-particle ground state. A peak at $\omega$ indicates that $\Delta_f^\dagger|N\rangle$ has an overlap with an $N+2$ particle eigenstate with energy $E_N+\omega$. Similarly, for $\omega < 0$ we obtain information on the $N-2$ particle state $\Delta_f|N\rangle$. One of the features of a superconducting state is that adding an additional pair *with the right pairing wavefunction* adds no energy to the system. In contrast, adding a single particle requires at least the gap energy, while trying to add a pair with the *wrong pairing symmetry* will create a mixture of excited states. Hence the signature of a superconducting state is a peak in $\bar{\chi}(\omega)$ at $\omega = 0$ when the right pairing symmetry is used. One would expect this peak to be present even above $T_c$; below $T_c$ one expects the area under the peak to diverge with the size of the system.

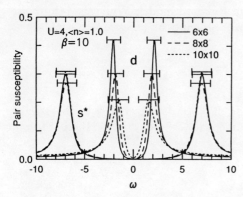

Fig. 5. Pair susceptibilities for the 2D Hubbard model at half-filling.

Figure 5 shows results for $\bar{\chi}_d(\omega)$ and $\bar{\chi}_{s^*}(\omega)$ for various lattice sizes at half-filling. At half-filling ($N = 16$) the system is antiferromagnetic, and $\bar{\chi}_d(\omega)$ reflects the presence of the insulating gap in $N(\omega)$. Hence there is no peak at $\omega = 0$; however, the peaks for the $d$-wave spectrum are at relatively low energies—roughly twice the single particle gap. From the $s^*$ spectra we see that adding a pair with that symmetry creates only a mixture of highly excited states. The results for $s^*$ show little finite-size effects. Further investigation will be necessary to tell if the apparently systematic shift in peak location for the $d$-wave spectra is statistically significant, given the limited resolution and the fact that the resolution is different for the different system sizes.

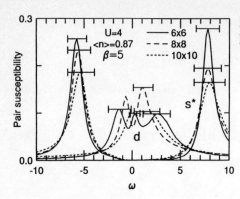

Fig. 6. Pair susceptibilities for the 2D Hubbard model away from half-filling.

The fillings where one would most expect superconductivity (perhaps $\langle n \rangle \sim 0.75$ to 0.95) correspond to fillings where fermion sign problems are worst. For these fillings, the lowest temperature which we could simulate accurately on the large lattices was $\beta = 5$. In the region $\langle n \rangle \sim 0.75 - 0.95$, where one can simulate only above a presumed $T_c$, the dynamic pair susceptibilities still provide valuable information, although they cannot prove the existence of superconductivity. Figure 6 shows results for a filling of 0.87 on $6 \times 6$, $8 \times 8$, and $10 \times 10$ lattices at $\beta = 5$. The $d$-wave spectra have peaks at low energies, but the area under the peaks does not vary appreciably with system size. This behavior is consistent with a superconducting system significantly above $T_c$. (One would not expect the peaks to appear precisely at $\omega = 0$ on a finite system.) Once again, the $s^*$ (and $s$, not shown) peaks lie at very high energies. Given the consistent behavior of the $s$ and $s^*$ spectra, it appears very unlikely that superconductivity with these symmetries occurs in the 2D single-band Hubbard model. While the results are consistent with $d$-wave superconductivity, it will probably be necessary to simulate at temperatures below $T_c$ to prove its existence.

## Acknowledgements

We wish to thank D.J. Scalapino, R.L. Sugar, R.T. Scalettar, and N.E. Bickers for helpful discussions and collaborations. The numerical work was done on the UC Irvine Convex and at the San Diego Supercomputer Center.

## References

1. D. Thirumvalai and B. Berne, *J. Chem. Phys.* **79**, 5029, (1983).

2. J. Hirsch, "Simulation of Magnetic Impurities in Metals", in *Quantum Monte Carlo Methods,* ed. M. Suzuki, Springer-Verlag, Heidelberg (1987).

3. H.-B. Schüttler and D.J. Scalapino, *Phys. Rev. Lett.* **55**, 1204, (1985); *Phys. Rev. B* **34**, 4744, (1986).

4. S.R. White, D.J. Scalapino, R.L. Sugar, and N.E. Bickers, *Phys. Rev. Lett.* **63**, 1523, (1989).

5. M. Jarrell and O. Biham, *Phys. Rev. Lett.* **63**, 1523, (1989).

6. R.N. Silver, D.S. Sivia, and J.E. Gubernatis, to be published.

7. S.R. White, UCI preprint.

8. S.R. White, D.J. Scalapino, R.L. Sugar, E.Y. Loh, J.E. Gubernatis, and R.T. Scalettar, *Phys. Rev. B* **40**, 506, (1989).

9. E.Y. Loh, J.E. Gubernatis, R.T. Scalettar, R.L. Sugar, and S.R. White, in *Proceedings of the Workshop on Interacting Electrons in Reduced Dimensions*, edited by D. Baeriswyl and D. Campbell (Plenum, New York, 1989).

10. E.Y. Loh, J.E. Gubernatis, R.T. Scalettar, S.R. White, D.J. Scalapino, and R.L. Sugar, to appear in *Phys. Rev. B*.

Part III

**High Performance
Computing**

# High Performance Computing in Academia: A Perspective

*W.B. McRae*

University Computing and Networking Services, The University of Georgia,
Athens, GA 30602, USA

Abstract. This paper reviews the experience of a task force of research faculty at the University of Georgia charged with evaluating alternative strategies and products to satisfy campus numerically intensive computing requirements. A survey of computer architectures examined by the task force is presented as well as the status of Fortran compiler technology on the systems examined. A discussion of some of the issues involved in effectively assessing system performance is also provided. The paper concludes with a discussion of alternative strategies considered by the task force to provide high performance computing resources.

## 1. Introduction

High performance computing has emerged as a powerful and indispensable aid to scientific and engineering research. Not only does this type of research resource bring an added dimension of understanding to existing problems, it also permits scientific exploration never before possible. Computationally intensive problems that were long considered intractable are now being solved. This tool allows exploration of phenomena not directly accessible to experiment in key scientific and engineering areas such as atmospheric science, astronomy, materials science, molecular biology, aerodynamics and elementary particle physics. When coupled with advances occurring in the graphical display and manipulation of numerical data, these resources have provided fundamental new insights into dynamical systems. Much of the emerging understanding of chaotic systems is a result of high performance computing studies.

The United States provided international leadership in the development and use of high performance computing from the inception of digital computing in the 1940s through the 1970s. By the early 1980s, this position of preeminence was beginning to erode. A number of studies focused on the causes for this decline with recommendations for redressing the erosion in high performance resources available, particularly in academia. One aspect of the problem was that federal support for large scale academic computing had all but disappeared in the late 1970s. Another aspect was the way campuses and researchers responded to the change in federal support patterns.

As an example of the way federal academic computing support changed, in the National Science Foundation the Institutional Computing Program and Regional Cooperative Computing Program were ended and the Office of Computing Activities was disbanded. Subsequently, NSF did not allocate funds directly for centralized computing at universities; funds were distributed by NSF disciplinary program areas, and researchers were encouraged to acquire equipment for their own use. This led to enormous growth in departmental and research project acquisitions of minicomputers in the 1970s. However, this approach also led to a situation in which central campus computing facilities were underdeveloped.

In this environment it was inevitable that campus computing planning and decision making became widely distributed. The computing center, academic departments, research institutes, the library, and individual faculty all became involved in the research computing enterprise, not only in the use of computing resources but also in their acquisition and management. In addition to this expansion of the number of people participating in the decision process, the range of available computing resources was also broadened, growing to encompass personal computers to large centralized systems. Many of the computing resources within this range were provided through the funding structures then in place, but universities were not able to meet all of the academic computing needs of their faculties, students and research staff. From the introduction of the current generation of vector supercomputers in the early 1970s through the early 1980s, no American university acquired a state-of-the-art supercomputer. In the view of many, the scientific capacity of the nation was weakened as a result.

In the early 1980s several pioneering universities, including Colorado State, Minnesota, Purdue and Georgia, moved to acquire supercomputers for support of academic research. Although the motives for these actions varied by institution, several considerations were commonplace. Each institution hoped by these initiatives to enhance their research competitiveness in attracting faculty and grant income. They all also believed that potential industrial subscribers would emerge to offset the substantial operating expense inherent in maintaining these systems in campus environments. Lastly, each expected that their early experience in using these systems would position them favorably to compete for federal funding that was anticipated to increase support for high performance computing in higher education.

The major federal funding initiative which did result was the NSF Advanced Scientific Computing Program. This program was launched in 1984. Although other federal agencies also supported high performance computing research initiatives, the NSF program had the most direct and dramatic effect on the access of large numbers of academic researchers to supercomputers. The program's goals were to "provide U.S. researchers with broad access to the most advanced computing capabilities, to introduce a new generation of students and young researchers to the potential of these advanced computing tools, and to stimulate innovation in the U.S. computer industry."

Phase I of NSF's program was designed to encourage the use of high performance computing at American universities by making available time and training on three (later expanded to six existing) supercomputers. Three of the original pioneering institutions were selected as Phase I centers. In 1984-5, under Phase II, NSF established three university based national supercomputer centers chosen from solicited proposals. The number of national centers was subsequently expanded to five when one was added later and a prototype center was converted to a national center.

These centers were intended to provide the high speed computing cycles, user services, training, specialized hardware, software and large memories needed to expand the basic and applied research problems for which computational solutions could be sought. Several other centers have also been established since this time, typically as state sponsored initiatives involving multiple universities. Notable examples are in Alabama, Ohio and North Carolina.

These various initiatives have begun to address the previous paucity of high performance computing services available to university researchers. We are now entering a new era, however, in satisfying the substantial demand which has been stimulated. The major challenge of this new climate will be maintaining the progress which has been made. Indications are that federal support, and NSF funds in particular, for high performance computing facilities will once again be diverted to other areas. The American high performance computing industry suffered a serious blow with the discontinuance by Control Data of their ETA product line. A competitive environment in the computer industry has always been advantageous for both the customer and supplier. The long heralded performance advantage of inexpensive massively parallel systems for large and complex codes is only slowly being realized. The commercial failure of many promising products of this type is testimony to this fact. Political forces and issues of national pride have also made considering Japanese product alternatives difficult and uncertain.

This paper reports on the experiences of a task force of research faculty at the University charged with evaluating alternative institutional strategies to meet this challenge. Access to high performance computing resources has become an indispensable research tool on our campus in a variety of disciplines. At the same time, our campus is calling for a much greater institutional investment in distributed information technology resources to support other legitimate university activities. Satisfying these seemingly disparate demands in a constrained funding environment is the challenge which all research universities will increasingly face.

## 2. Architectures and Compilers

Numerous classifications or taxonomies for computer architectures have been presented [1,2,3] particularly as they relate to implementation variations for parallelism in the organization of computational elements and data memory. The most firmly established classification scheme is that due to Flynn [4] because of its historical originality and simplicity. He divided computer architectures into four broad categories. These were:

  i. SISD - single instruction stream, single data stream
  ii. SIMD - single instruction stream, multiple data stream
  iii. MIMD - multiple instruction stream, multiple data stream
  iv. MISD - multiple instruction stream, single data stream

SISD describes conventional Von Neumann uniprocessors which contain one instruction sequencing unit and are capable of retrieving and manipulating one data item (or a pair of data items) at a time. In SIMD architectures, one central execution unit broadcasts a specific instruction to multiple processing elements (PEs). The PEs can operate on different data items, but can only execute the particular instruction issued or do nothing in a given cycle. MIMD implementations are an interconnection of independent execution units, each capable of processing different data items fetched from shared or local memories. Most commercial multiprocessing systems today can be classified as MIMD systems. No known implementations of the fourth Flynn category, MISD, have been seriously pursued. Typically, it is listed for the sake of completeness and symmetry.

Although these categories give a helpful coarse division, many current machines exhibit features of more than one category. For example, large CRAY systems have multiple processors (MIMD), but each processor uses segmented instruction execution units to provide vector operations on multiple data items (SIMD). Moreover, some current architectures such as Long or Very Long Instruction Word (LIW or VLIW) machines and data flow systems do not fit easily into this classification at all.

A different classification scheme has been proposed by Serlin [5]. Although his scheme is less precise and primarily empirical, it has the strength of adequately describing the majority of products available today. This is not a pedestrian quality in an environment which sees new products announced (and vacated) monthly with claims of superior architectural advantage for high performance scientific computing. In the Serlin taxonomy, five different implementations of parallelism are identified. These are:

i. VP - vector processing
ii. (V)LIW - (very) long instruction word processing
iii. SIMD - (per Flynn)
iv. MIMD - (per Flynn)
v. DFA - data flow architectures.

The first two categories refer to forms of parallelism that exist within a given processor. The next two categories describe parallel systems using multiple processing elements. The last category is an implementation completely outside the Von Neumann frame of reference. The major features of each of these classifications is reviewed below as an aid in following the concepts presented in the balance of this paper.

**Vector processing.** One of the most common commercial manifestations of parallel processing available today is vector processing. It is present in all current supercomputers from CRAY, NEC, Fujitsu and Hitachi; and in the mini-supercomputers from Alliant, Convex and SCS.

Vector processing makes use of what are called "pipelined" or "segmented" arithmetic instruction processing units. The pipeline consists of separate stages or segments which perform the primitive operations involved in a particular floating point arithmetic operation such as addition on consecutive data elements, i.e., vector elements. Floating point numbers and arithmetic are the means by which real number data is represented and manipulated in computer operations (as opposed to integer and character data). It is similar in form to the traditional scientific representation of numbers using base and exponent notation. Examples of the primitive operations which can be performed in a vector pipeline stage are exponent and decimal (binary) point normalization followed by exponent arithmetic and mantissa arithmetic. The goal in vector processing is to implement a sufficient number of primitive function stages in a pipeline so that each can complete processing in one basic clock cycle. If this is achieved, then after vector elements equal to the number of stages are present in the pipeline, a final computed result can be issued each clock cycle. By replicating pipelines and synchronizing the distribution of vector elements across these pipelines, more than one result per clock cycle also can be achieved. The number of cycles necessary to achieve the first computed result from a vector pipeline is often referred to as the vector pipeline length or start-up time.

**Long and Very Long Instruction Word Machines.** These machines consist of a single instruction sequencing unit, but multiple, different functional units. Each long instruction fetched from memory can contain a sub-instruction for each of the functional units. The key to increased performance in this type of machine is an intelligent compiler which is capable of achieving parallelism by ordering conventional serial code into long instruction words that keep as many functional units busy in each clock cycle as possible. Typical implementations include from 8 to 16 functional units in a system. Proponents of VLIW-LIW machines claim that these architectures can extract parallelism from code segments that both vector and multiprocessor systems cannot. Current examples of such architectures include the Multiflow Trace products and FPS M series. Earlier systems no longer manufactured include the Cydrome system and CDC CyberPlus.

**SIMD.** SIMD machines also have been referred to as data parallel computers. They are characterized by a single instruction sequencing unit which issues the same instruction to multiple processing elements. Each processing element then carries out that operation on different data elements simultaneously. Local data memories for each processor are a requirement of this architecture. Some argue that this simplicity leads to less parallel programming flexibility although SIMD proponents assert that most algorithms that benefit from parallelism map to the SIMD architecture very well. The Connection Machine from Thinking Machines Corp. is a currently successful commercialization of SIMD architecture. The AMT DAP is an even earlier SIMD implementation as were the venerable Illiac IV and Loral/Goodyear MPP.

**MIMD.** MIMD implementations incorporate multiple independent processors in various arrangements. Conventional multiprocessing systems, or "coarse grain" parallel systems are designed to improve system throughput by concurrently servicing multiple, unrelated tasks. Typically, when separate central memories are available to each processor in these configurations, the system is called "loosely-coupled." One successful design of this type is from Tandem Computers, which offers a loosely-coupled MIMD for fault tolerant applications. Other loosely-coupled, conventional multiprocessor designs are available from DEC, NCR and others. A shared central memory characterizes "tightly-coupled" multi-processor MIMD systems. IBM is a significant player in this MIMD arena with its 3090 multiprocessor series which can also be equipped with adjunct vector processing units to supplement each basic processor unit. Other companies supplying tightly-coupled MIMD systems include Alliant, Convex, Concurrent, Pyramid and Sequent.

MIMD systems can also be designed for "fine-grain" parallel processing, i.e., concurrent instruction execution within the same computational task. A very significant implementation of this type is the hypercube architecture. Hypercubes are attractive in that they employ a nearest neighbor inter-processor communication scheme which makes very large configurations practical. The number of directly connected nodes prescribes the dimension of the hypercube. For example, a system with 32 nodes is a 5 dimensional hypercube with each node directly connected to five nearest neighbor nodes. Memory is local to each node with interprocess communication accomplished by message passing between consecutive nearest neighbors, thereby eliminating the bus contention that can occur in shared memory systems. Systems within excess of 1000 processors are available and are often called "massively parallel" systems. Intel Scientific Computers and NCUBE are the leading commercial suppliers of hypercube products.

Massively parallel systems can also be built using a multistage network as the interconnect mechanism. These are normally implemented using staged, simple crossbar switches to achieve complete logical connectivity with fewer physical links than would be required for full physical connectivity. Proponents claim that this approach eliminates the store-and-forward penalty inherent with inter-processor communication in hypercube designs. Currently, the only commercial product in this class is the BBN Butterfly. The IBM RP3 research project uses a similar interconnect.

**Dataflow.** These systems represent an interesting departure from the Von Neumann model. In a Von Neumann machine, the programmer, either directly or through a compiler, specifies the sequence of instructions to be executed. In a dataflow machine, computation is data driven: an operation can be initiated as soon as all of its input operands have been computed and a functional unit of the appropriate type is available. The key difficulty with dataflow systems is that they require new programming languages and algorithmic designs. MIT is seeking industry support for a commercial realization of dataflow ideas called the Monsoon architecture.

**Compiler Characteristics.** The computer architectures described in the preceding section are theoretically capable of very high execution rates. Unfortunately, these peak performance rates are seldom, if ever, achieved in practice. Typically the performance of the most powerful vector, multiprocessor systems with actual large scientific codes does not exceed 20% of the advertised peak performance potential. These differences derive from several factors. First of all, peak performances typically presume optimal utilization of all the instruction processing capability inherent in a particular architectural implementation. As a result, the instruction sequence required to exploit this potential is very synthetic and not characteristic of most mathematical algorithms. Because of this, peak performances quoted by manufacturers have often been referred to more accurately as the performance level that can never be exceeded for a particular machine. Of more fundamental importance, however, is the requirement that for a program to effectively utilize available high performance computing architectures it must contain inherent parallelism and be expressed in a form that a compiler can recognize and exploit. Algorithmic expression in scientific computing necessarily requires considering the suitability of the FORTRAN programming language and the various compilers for this language as a tool for capturing parallelism. This requirement is best reflected in the classic statement attributed to A. Perlis that "FORTRAN is not a flower, but a weed. It is hardy, occasionally blooms, and grows in every computer."

The advent of vector processors and multiprocessors is only now beginning to influence FORTRAN and its various dialects. Vector statements have been standardized to the extent that they will be part of the next FORTRAN standard [6]. No consensus yet has been reached with regard to language extensions for multiprocessing. This circumstance is reflected in the fact that most compilers available today for high performance computer systems presume standard FORTRAN 77 as input with the parallelization accomplished implicitly by the compiler. To insure that this implicit optimization is accomplished more effectively, many compilers provide the option of programmer supplied directives to assist in detecting code segments eligible for vectorization or multi-tasking. One exception to this general practice has been the CDC CYBER 205 FORTRAN which supports explicit vector statements. A number of tools that perform program restructuring for vectorization and parallelization which report back to the programmer also have been developed to assist in program optimization. Examples are Parafrase [7] developed at the University of Illinois, PFC [8] at Rice University, and VAST [9] and FORGE which have been developed by the Pacific-Sierra Corp.

Comparative evaluations of the optimizing efficiency of the numerous FORTRAN compilers available today are now beginning to appear [10]. The most exhaustive study of this type performed is that carried out by the Argonne National Laboratory [11]. The objective of the Argonne research is to test the efficiency of vectorizing compilers in four broad areas: dependence analysis, vectorization, idiom recognition, and language completeness. Dependence analysis refers to process of collecting information about array subscripts and subsequent testing for memory overlaps between pairs of variables. Vectorization reflects the ability of the compiler to transform FORTRAN DO loop constructs into explicit vector instructions. Idiom recognition refers to the identification of particular program forms that have presumably faster special implementations such as vector sums and dot products. Language completeness tests how effectively a compiler can recognize vectorization implicit in FORTRAN constructs other than simple DO loops containing only primitive floating point and integer assignment statements. The methodology followed by the Argonne group was to compile a set of loops written in standard serial FORTRAN which tested specific features of vectorizing compilers. These loops were provided to vendors who were asked to compile the loops without making any changes, using instead only those compiler options that provided for automatic vectorization. The output listings were then evaluated by Argonne to determine which loops had been fully or partially vectorized. The results for 100 loops which were compiled successfully by all participants and which tested different features were then published [11].

The major conclusion from this study is that no currently available compiler is capable of fully vectorizing serial FORTRAN, let alone other forms of intrinsic parallelism. On the average, slightly more than one half of the possible loops were fully vectorized in this test. The best result reported for full vectorization is 69%. Clearly, extensions to FORTRAN must be achieved if the performance potential of the current and emerging high performance computing systems is to be realized.

Three of the extensions proposed for the long gestating FORTRAN 8x standard which will be briefly described here focus on more effective vector programming [12]. These are triplet expressions, IDENTIFY statements, and WHERE statements. The triplet notation consists of three expressions separated by colons. These expressions indicate the range of execution of this statement and correspond to the beginning, end and stride for this range--as in a FORTRAN DO loop. If the beginning of a triplet is omitted, it is assumed to be the beginning of the array; if the end is omitted, it is assumed to be the end of the array. Unless otherwise specified, the stride is assumed to be 1, and it may be omitted along with its separating colon. This notation makes it possible to assign a section of one array to a section of another array or array section expression. Array sections also can be specified for multidimensional arrays with one triplet for each dimension of the array, and triplet notation may be mixed with the normal array index notation. The only restriction is that two sections appearing together in the same statement must have an equal number of elements in each corresponding dimension. An example of a one-dimensional array assignment statement using triplet notation is:

A(1:3:2) = B(2:4:2) * C(3:8:4)

which is equivalent to

A(1) = B(2) * C(3)A(3) = B(4) * C(7).

Triplets are not sufficient to express all vector operations. Specifically, conditional operations cannot be performed with triplets. The WHERE statement provides for conditional vector assignment, or what is sometimes called masked assignment. The WHERE statement functions by first evaluating a logical array expression. The statements in the body of the WHERE are executed for each index value evaluated to be true. The body of the WHERE statement contains array assignment statements for which the component array variables must be conformable to the logical array expression. If an OTHERWISE block is present, the array assignment statements of this block will be performed for every corresponding element of the logical array expression of the WHERE statement whose value is false. For example, to inhibit the taking of SQRT for negative values, the following code sequence may be used:

WHERE(X>=0)
Z = 4.0 * Y * SQRT(X)

OTHERWISE
Z = 0.0
END WHERE.

The IDENTIFY statement permits array operations which cannot be achieved using triplets and WHERE statements. An example of this occurs in operations which have a regular stride in physical memory but an irregular stride in array indices. Assigning the diagonal elements of a two-dimensional array is an illustration. The IDENTIFY statement allows aliasing to a part of an array so that such operations can be performed. The following code illustrates this facility:

REAL A(10,10)
IDENTIFY (DIAG(I) = A(I,I),I = 1,10)
DIAG(:) = 1.

No comparable standardization for FORTRAN extensions has been attained for multi-tasking constructs to utilize multiprocessor MIMD architectures. Multiprocessor systems present greater challenges in programming and compiling than vector machines. In general, vector programming is still serial programming where optimization involves local decisions within an algorithm. Effective program design and compilation for multiprocessors requires global consideration of the program and data dependencies. Additionally, utilization of a multiprocessor often requires careful consideration of the particular machine architecture employed because of the variability in implementations of MIMD designs. For example, shared memory architectures impose different constraints than local memory systems. As a result, the lack of standardization is understandable.

Amongst the numerous vendor dialects which have evolved, two basic multi-tasking methods have emerged to increase the performance of a single program: macro-tasking and micro-tasking. Macro-tasking normally refers to the segmentation of a problem into separate components called tasks that can execute more or less independently on the multiprocessors. Initiation and synchronization of these tasks is usually achieved through fork and join constructs [13]. Creating a new task is generally a very expensive operation. The useful work performed by the task must be sufficiently large to compensate for this overhead if macro-tasking is to be efficient. Micro-tasking exploits parallelism at a more local level than macro-tasking. Micro-tasking is often used to execute iterations of a loop in parallel. In addition to loop parallelism, speedups can be achieved by overlapping segments of sequential code. This operation is called low level spreading if the spreading is done on a statement-by-statement basis; it is called high-level spreading if the spreading is performed upon large instruction streams [14]. Micro-tasking is efficient on such small execution units because it does not incur the large task creation overhead of macro-tasking. The variety of multi-tasking libraries and constructs available are increasingly leading to legitimate complaints that parallel programs are not transportable and difficult to write. As a result, earlier versions of automatic parallelizers are beginning to appear. Notable examples are recent versions of IBM VS FORTRAN and Alliant FORTRAN. Generic environments such as LINDA [15] have been created which provide uniform parallel programming language specifications that can be implemented on a variety of multiprocessor and distributed systems. In these developments are the seeds for the eventual standardization of parallel programming constructs which are just now being realized for vector systems.

## 3. Benchmarks

Comparative performance measurements for high performance computing systems have received a great deal of attention recently. This interest primarily reflects the extremely competitive and dynamic market environment which exists for these systems. The performance capability of very large scale systems relative to Japanese products has even been linked to national pride and our future technological competitiveness by some [16]. Because of the price of many of these systems and the controversy their acquisition can generate, developing realistic measures for their performance benefit can be critically important in justifying procurement decisions and minimizing investment risk.

Several programs or sets of programs have evolved as industry standards for comparative performance timings. The best known amongst these are the LINPACK timings [17] produced by the Argonne National Laboratory and the Livermore Kernels. The LINPACK results measure the performance of different computer systems in solving a dense system of linear equations using the LINPACK FORTRAN library routines originally developed at Argonne for this purpose. LINPACK measurements have been carried out on a very large number of computer systems under rigorous conditions established by Argonne. The results are regularly updated to reflect newly introduced computer products in publications available to the public. The Livermore Kernels are prototypical algorithms extracted from larger codes which are characteristic of the historical work load at this laboratory. They have traditionally been used by Livermore to evaluate subsequent generations of high performance vector processors which this laboratory uses extensively. Performance data for programs specific to particular discipline areas such as GAUSSIAN in computational chemistry and NASTRAN in structural engineering is also increasingly cited by manufacturers targeting specific engineering and scientific audiences for their products.

Enormous confusion and differences of opinion surround the interpretation of the timing results obtained from these traditional measurement programs. Disagreement exists over the interpretation of the units of performance used, the degree of optimization permitted, sizes of problems studied, and the methods used to analyze the data obtained. To bring some clarity to this situation, several associations including both suppliers and consumers have recently been formed with the goal of establishing uniform performance evaluation criteria and procedures for high performance computing systems [18]. Notable amongst these have been the Systems Performance Evaluation Cooperative (SPEC) and the Perfect Club (Performance Evaluation for Cost Effective Transformation). Although these developments bode well for the future prospective buyer of high performance computing systems, they will not relieve customers from carefully evaluating competing products relative to their own peculiar work load setting and support needs.

We do not propose in this section of our paper to dictate to others about meeting this challenge in their own environments. Instead, we will briefly review the procedures which were followed by a task force of research faculty at The University of Georgia in obtaining comparative performance information.

An initial assessment of alternative computing architectures and products potentially available to meet our campus high performance computing need was conducted by inviting a number of manufacturer representatives to address the task force. An extensive and standard list of information items was developed to which each of the prospective suppliers was required to respond. Not every potential supplier was evaluated. Instead, the task force solicited presentations on what were considered to be representative products for generic categories of computer system options. The intent was not to evaluate exhaustively all products available but rather to evaluate the suitability of the various architectural and performance options available in arriving at a recommended strategy for the campus. The major categories of products considered were true "supercomputers," "mini-supercomputers," and high performance workstations.

A total of eleven different product presentations were provided to the task force. Table 1 lists the specific manufacturers and their products evaluated.

### Table 1

#### Products Evaluated

| Parallelism | Company | Product |
|---|---|---|
|  |  | *True "Supercomputers"* |
| VP & MIMD | Cray Research | Cray Y-MP |
| VP & MIMD | IBM Inc. | IBM 3090-600 |
| VP & MIMD | NEC | HSNX SX-X |
|  |  | HSNX SX-2 |
|  |  | *"Mini-Superscomputers"* |
| VP & MIMD | Alliant Computers Sys. Corp | Alliant Sunrise |
| VP & MIMD | Convex Computers Inc. | Convex C2XX Series |
| VP & MIMD | Digital Equipment Corp. | DEC Aquarius |
| VLIW | Multiflow Computer Corp. | Trace 400 Series |
| SIMD | Thinking Machines, Corp. | Connection Machine |
| MIMD | Sequent Computing Corp. | Symmetry Series |
|  |  | *High Performance Workstations* |
| VP | Ardent Computers, Inc. | Titan P3 Series |
| VP | Control Data Corp. | CDC 910-920 Series |

Next, the task force profiled the campus research computing work load that could benefit from high performance computing resources. Three general areas were identified in this process: computational chemistry, computational physics and statistical analysis supporting social science research. Prototypical codes from each of these areas in current use by faculty were then collected. Program codes were then selected from those provided that were written in standard FORTRAN to insure machine independence to the extent possible. This requirement also complemented the task force's desire to measure the performance advantages of the various architectures evaluated for existing programs without modification. To insure that the timing results obtained reflected meaningful problems, cases were selected that required at lease five minutes of central processor time in an IBM 3090 environment. Vendor participants were asked to measure the performance of the codes supplied without change initially and then in a separate execution to exploit their particular machine architecture and programming features.

The results obtained from this performance assessment were generally consistent with the task force's expectations. They showed first of all that only the "true supercomputers" tested were capable of providing substantial performance advantage without significant code conversion and optimization. Currently available "minisupercomputer" products appeared to manifest inadequate performance to satisfy the campus requirements unless multiple systems were contemplated or the scaled performance advantages claimed for future products considered. The task force also concluded from the results that high performance workstations could not meet the campus need.

## 4. Access Strategies

The faculty task force charged with developing alternative strategies to meet the numerically intensive research computing requirements on The University of Georgia campus recommended acquiring a centrally managed entry level high performance vector, multiprocessor system. They further recommended that this entry level system should be viewed as a platform for developing an ultimate shared large scale computing resource for the several research institutions within the University System of Georgia. In reaching this conclusion, the task force considered many of the issues which normally surround the debate in academic institutions over high performance computing resources. Amongst these were questions of centralized versus decentralized facilities; the suitability of specialized versus general purpose systems; and lastly the merits of local campus access to these resources as opposed to remote access. A summary of some of the conclusions reached by this faculty task force are presented here for your information. Alternative conclusions could well be reached in different academic environments. In fact, these views are not unanimously accepted by other faculty at our institution. The conclusions reached, however, do reflect judgments reached after careful and extensive consideration by the faculty involved in this study of the full spectrum of complexities involved in these issues.

The task force recommended a centralized resource as opposed to decentralized facilities distributed throughout the campus for a number of reasons. They concluded that economies of scale still apply to acquisition of high performance computing resources in institutions, both with respect to capital and support costs. This consideration is particularly relevant when subsequent and inevitable expansion of the resource must be addressed. A number of "mini-supercomputers" were carefully examined by the task force in reaching this conclusion. The group's observation was that the measured performance for these systems was simply inadequate without appealing to future products to meet existing campus work load requirements. The task force also believed that determining which research groups would be the beneficiaries of any institutional investment in decentralized systems would be very difficult. The perception that inequities existed in allocation of computing resources on campus would be unavoidable. The group felt that providing computing resources to the increasingly common "computational scientists" among the faculty across various disciplines could be better managed in the central environment.

Another consideration was that the task force believed that the synergistic advantages of collaborative research and the sharing of numerical methods could best be achieved through a single, common high performance computing resource.

Acquisition of specialized systems with architectures and programming environments which could be optimized for algorithms characteristic of specific research methods was considered very seriously by the task force. Implicit in this strategy was the requirement that multiple different systems would have to be acquired. Ultimately this approach was considered inappropriate for many of the same reasons that decentralized resources were not recommended. Several other disadvantages for this strategy were also raised. Distributing a dynamic work load across these systems to insure optimal utilization would be difficult because of their dissimilar programming environments. Programs optimized for one environment would not process efficiently, if at all, in a different architectural setting. This circumstance could potentially result in more frequent capacity upgrades for individual systems than would be necessary if the resources could be effectively shared. Separate maintenance service costs and support staff expense would have to be incurred for each system acquired as well. Another concern of the task force with the systems which would plausibly be acquired under this strategy was their throughput capacity. Typically these systems incorporate parallel architectures which realize their attractive price/performance characteristics by distributing a particular computing task across multiple relatively slow (hence inexpensive) instruction processing elements. This can be of significant performance advantage when such a system is dedicated to a single task where all of the processing elements can be assigned to the task. In a multiple user environment, however, the number of processing elements available to each task is reduced and the time to completion for an individual task can be dramatically extended.

Use of the NSF centers as remotely located suppliers of high performance computing resources also was carefully examined by the task force. These centers were considered plausible for evaluating alternative supercomputer environments and evaluating whether a particular research application can benefit from such facilities, but reliance on NSF centers for established and continuing resource intensive research programs was not considered feasible by the task force. Task force members who had used NSF facilities in the past indicated skepticism relative to the availability of meaningful amounts of time. It was felt that these centers suffer from serious over subscription of users as an administrative policy in order to dramatize demands to the NSF for more resources. A user population in excess 10,000 [19] is now claimed across the five Centers. If this is, in fact, the case and time were to be allocated uniformly to this user community, less than five hours per year would be available to each individual user. The inadequacy of the available network bandwidth to access these Centers from The University of Georgia through the SURAnet network was also cited as a serious problem with their use for the file transfer activities characteristic of supercomputer applications. The SURAnet backbone links are 56 kbps DDS telephone circuits which connect to NSFnet T1 (1.54 mbps) trunks through gateway processors. These links are simply inadequate to support large numbers of users other than in a simple timesharing mode. Compounding this difficulty is the thirty percent per month growth rate for NSFnet trunk traffic. In recognition of this problem, NSF is expected to begin phasing in 45 mbps trunk data channels in fiscal year 1990 if funding and technology permits. Perhaps more critical than either of these difficulties in arguing against reliance on NSF centers as a solution to the campus high performance computing requirements is the stated intention by NSF to convert funding for these Centers to a "market mechanism." [20] Under this approach, direct NSF support for these Centers would be substantially reduced with the funding released made available to the various NSF scientific discipline programs which would likely use these facilities. The most optimistic proponents for this change fully expect that the current free use of Center resources will be discontinued under a "market mechanism" at a minimum. Less sanguine observers believe that it may mean the end to the NSF Centers.

In conclusion, it is worth noting that the challenge of developing the most appropriate strategy to meet the demand for high performance computing resources on our campus is still being debated. Even though the task force charged with developing proposals for this purpose believes its recommendation was sound, the very difficult administrative decision to allocate funds to this goal rather than to other equally pressing needs remains to be made.

## REFERENCES

1. R. E. Anderson, In *SuperComputing '89 Proceedings*, November 13-17 (1989).

2. J. J. Dongarra & I. S. Duff, *Advanced architecture computers* (MCS-TM-57). Illinois: Argonne National Laboratory, October (1985).

3. P. H. Enslow, Computing Surveys **9**, 1 (1977).

4. M. J. Flynn, IEEE Transactions on Computers **C-21**, 9 (1972).

5. O. Serlin, Parallel Processing Reports **I20**, January (1989a).

6. ANSI FORTRAN 8x, x3J3/S8 (x3.9 - 198x), American National Standards Institute (1986).

7. D. Kuck, R. Kuhn, B. Leasure & M. Wolfe, The structure of an advanced vectorizer for pipeline processors. Paper presented at the 4th International Computer Software and Appllctions Conference, October (1980).

8. R. Allen & K. Kennedy, *PFC: A program to convert FORTRAN to parallel form* (Tech. Report MASC-TR 82-6). Texas: Rice University, Department of Mathematical Sciences, March (1982).

9. B. Brode, VAST - A vectorization tool for the CYBER 205. Symposium on CYBER 205 Applications, August (1982).

10. A. H. Karp, & R. G. Babb, *A comparison of some dialects of parallel FORTRAN* (Technical Report G320-3511). CA: IBM Palo Alto Scientific Center (1988). Applications Conference, October (1980).

11. D. Callahan, J. Dongarra & D. Levine, *Vectorizing compilers: A test Suite and Results* (MCS-TM-109). Illinois: Argonne National Laboratory, March (1988).

12. M. D. Guzzi, D. A. Padua, J. P. Hoeflinger & D. H. Lawrie, In *Supercomputing '88 Proceedings*, November, 14-18 (1988).

13. J. Dennis & E. VanHorn, Communications of the ACM **9**, 3 (1966).

14. A. Veidenbaum, Compiler optimizations and architecture design issues for multiprocessors (CSRD Doc. No. 520). Urbana: University of Illinois, Center for Supercomputing Research and Development (1985).

15. S. Ahuja, Carriero, N. & Gelernter, IEEE Computer **19**, 8 (1986).

16. J. Worlton, Datamation September (1984).

17. J. J. Dongarra, *Performance of various computers using standard linear equations software in a Fortran environment* (MCS-TM-23). Illinois: Argonne National Laboratory, June (1989).

18. O. Serlin, Performance standards emerge. *Parallel Processing Reports* **I31**, December (1989b).

19. M. Sun, Science **245** (1989).

20. O. Serlin, Parallel Processing Reports **I24**, May (1989c).

Part IV

**Contributed Papers**

# Finite-Size Scaling Study of the Simple Cubic Three-State Potts Glass

M. Scheucher[1], J.D. Reger[1], K. Binder[1], and A.P. Young[2]

[1]Institut für Physik, Universität Mainz, Staudingerweg 7,
W-6500 Mainz, Fed. Rep. of Germany
[2]Physics Department, University of California, Santa Cruz, CA 95064, USA

During the last few years the Potts glass model has attracted more and more attention. It is considered as a first step towards modelling the phase transition of structural and orientational glasses. A mean-field approach /1/ predicts a low temperature behavior completely different from what is known from Ising spin glasses /2/. But short range models differ markedly from mean-field-predictions. So it is natural to ask, how the short range Potts glass behaves. Especially the question of the lower critical dimension $d_l$ is important, below which a finite temperature transition ceases to occur. We tried to answer this by combining Monte-Carlo simulations with a finite-size scaling analysis. The model is defined by the following Hamiltonian

$$H = -\sum_{<i,j>} J_{ij}\delta_{n_i,n_j} \quad ; \quad n_i \in \{1, 2, \ldots, p\} , \quad (1)$$

where the sum is over all nearest neighbor pairs of the simple cubic lattice. The couplings are Gaussian with zero mean and variance one, which merely sets the temperature scale. We further restrict to $p = 3$.

In these units the transition temperature in the mean-field approximation is $T_c^{MF} = \sqrt{z}/p$, where $z$ is the coordination number of the lattice. In our case of $z = 6$ and $p = 3$ this gives $T_c^{MF} = 0.8165$.

In the following it is convenient to use the "simplex" representation for the Potts spins, i.e. introduce p unit vectors embedded in $(p-1)$ dimensional space : $\vec{S}^\alpha, \alpha = 1, \ldots, p$, with the property

$$\vec{S}^\alpha \cdot \vec{S}^\beta = (\delta_{\alpha,\beta} - \frac{1}{p})\frac{p}{p-1} .$$

Each of the $p$ states is then represented by one of the vectors. We then associate an appropriate vector $\vec{s}_i$ with each lattice site $i$, and set it to the corresponding basis-vector, i.e. if site $i$ is in state $n_i$, we set $\vec{s}_i = \vec{S}^{n_i}$. Thus the Hamiltonian reads

$$H = -\sum_{<i,j>} J_{ij}\vec{s}_i \cdot \vec{s}_j .$$

As in earlier works on spin glasses we study the behavior of the distribution of the order parameter /3/. The order parameter is defined in terms of the overlap between 2 real replicas of the system. Here replica means 2 statistically independent copies of the system, with an identical realization of the bonds and no coupling between them. The mutual overlap at any instant of time is then given by

$$q^{\mu\nu} = \frac{1}{N}\sum_{i=1}^{N} s_{i,1}^\mu \cdot s_{i,2}^\nu , \quad (2)$$

where $\mu, \nu$ label the components of the state vectors and "1,2" refer to the different replicas.

In order to sample the phase space most efficiently and to keep the relaxation times as short as possible, the order parameter has to be invariant under the symmetries of the Hamiltonian. This is achieved by defining

$$q = \sqrt{\overline{\sum_{\mu,\nu}(q^{\mu\nu})^2}} \quad , \tag{3}$$

which is clearly invariant under the permutation of the states in either replica. For systems of linear sizes ranging from $L = 3$ to $L = 8$ we ran Monte-Carlo simulations and measured the order parameter distribution $P(q)$.

Because of the dramatically increasing relaxation times our simulations were limited to rather small lattice sizes. For the largest system $4 \cdot 10^6$ MCS were needed at the lowest temperature (T=0.3), where we discarded $2 \cdot 10^6$ MCS for equilibration. Between 400 and 500 different samples were taken to average over the disorder.

We now focus on the finite size behavior of the moments

$$<q^n> = \int q^n P(q) dq \quad .$$

Note that $<q^2>$ is simply related to the spin glass susceptibility $\chi_{SG}$ by

$$\chi_{SG} = N<q^2> = \frac{1}{N}\sum_{i,j}[<\vec{s}_i \cdot \vec{s}_j>_T^2]_{av} \tag{4}$$

where $<\ldots>_T$ denotes the thermal average for a given bond configuration and $[\ldots]_{av}$ the average over the disorder. The finite-size scaling ansatz for the moments is

$$<q_L^n> = L^{-n(d-2+\eta)/2} f_n(L/\xi_{SG}(T)) \tag{5}$$

with $\eta$ the usual critical exponent describing the decay of the spatial correlations at $T_c$ and $\xi_{SG}(T)$ the correlation length.

This gives the following expression for the susceptibility

$$\chi_{SG}(T) = L^{2-\eta} f_2(L/\xi_{SG}(T)) \quad . \tag{6}$$

As written, eqs (5,6) should be valid not only for a system below or above its lower critical dimension, where $\xi_{SG}(T)$ diverges with a power of $T$, but also for a system right at its lower critical dimension, where $\xi_{SG}(T)$ diverges exponentially, i.e.

$$\xi_{SG} \sim \begin{cases} (T-T_c)^{-\nu} & d > d_l \\ T^{-\nu} & d < d_l \\ \exp(CT^{-\sigma}) & d = d_l \end{cases} \tag{7}$$

For the latter case we allow $\sigma$ to be different from one, in contrast to pure systems, and in fact, a simple scaling argument by McMillan /4/ predicts $\sigma = 2$.

Note that in all situations

$$\chi_{SG} \sim \xi_{SG}^{2-\eta} \quad . \tag{8}$$

As implied by eq.(5) it is convenient to study the "renormalized coupling" $g_L$, given by

$$g_L = 3 - 2\frac{<q_L^4>}{<q_L^2>^2} = \tilde{g}(L/\xi_{SG}(T)) \quad , \tag{9}$$

which is only a function of the scaled variable $L/\xi_{SG}(T)$, since the L-dependent prefactors in front of the scaling functions cancel.

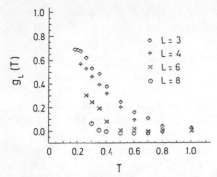

Figure 1: Reduced cumulant $g_L$ plotted vs. $T$ for different lattice sizes. These results were obtained from Monte Carlo runs with up to $4 \cdot 10^6$ MCS (for $L = 8$ at $T = 0.3$) and typically an average over $400 - 500$ bond configurations was performed. Note that $g_L$ is defined such that $g_{L \to \infty} = 0$ in the disordered phase, $g_{L \to \infty} = 1$ in a phase with nonzero order parameter, while curves for $g_L$ for different sizes should intersect at a critical point.

In our normalization $g_L \to 0$ as $L \to \infty$ in the high temperature phase, $T > T_c$, because the components of $q^{\mu\nu}$ have independent Gaussian fluctuations. Furthermore $g_L \to 1$ as $L \to \infty$ in the ordered phase, since $<q^4> = <q^2>^2$.
If there is a finite $T_c$ then $g_L$ should have an additional nontrivial fixed point and the curves for different $L$ should intersect at $T_c$.
However, as one sees in fig.1, there is no indication of a phase transition down to temperatures of $T = 0.15$, so if there should be a finite $T_c$ then it is incredibly small compared with the mean-field value of $T_c^{MF} = 0.8165$.
It would be a coincidence if $T_c$ were finite yet extremely small, so it is more likely that $T_c$ is equal to zero, since this occurs whenever $d \leq d_l$.
In fig.2 we present our data for $\chi_{SG}$. The log-log plot in fig.2a shows that there is a distinct curvature in the regime for $T > 0.5$, where the data has (nearly) settled down to a size-independent value. Hence if $T_c = 0$ either we have an exponential divergence as $T \to 0$ or if $\chi_{SG}$ diverges with a power, then the asymptotic regime where this occurs has not yet been reached for $T = 0.5$.

On the other hand the log-log plot of $\ln \chi_{SG}$ against $T$ shown in fig.2b is consistent with the data settling down on the curve $\ln \chi_{SG} \sim T^{-\sigma}$ with $\sigma = 2$.
To find further evidence for $d_l = 3$ we used a procedure which does not enforce any of the asymptotic forms of $\xi_{SG}$ in eq.(7), but rather tries to extract it from the data itself. To do so we note that $\chi_{SG}(L,T)/\chi_{SG}(T)$ depends only on the scaled variable $L/\xi_{SG}(T)$, where $\chi_{SG}(T)$ is the value of the susceptibility of the infinite lattice. For each temperature we estimate $\chi_{SG}(T)$ by extrapolation and determine a characteristic length $\ell(T)$ by demanding that the data for $\chi_{SG}(L,T)/\chi_{SG}(T)$ falls onto a single curve when plotted against $L/\ell(T)$. The $\ell(T)$ are freely adjustable parameters. From finite-size scaling we know that $\ell(T) \sim \xi_{SG}(T)$ without any particular assumption on the asymptotic form.
The scaling plot of $\chi_{SG}(L,T)/\chi_{SG}(T)$ is shown in fig.3, while fig.4a and fig.4b are log-log plots of $\ln(\chi_{SG}(T))$ vs. $T$ and $\ln(\ell(T))$ vs. $T$ respectively. Both being nicely consistent with the behavior expected for a system right at its lower critical dimension.
From eq. (8) it follows that the dependence of $\chi_{SG}(T)$ on $\xi_{SG}(T)$ gives the exponent $(2 - \eta)$. The data shown in fig.4 gives $(2 - \eta) = 1.5$ and thus $\eta = 0.5$. This is to be contrasted with $\eta = -1$ as expected for a system with $T_c = 0$ and a non-degenerate ground state. Since our value of $\eta$ is very different from this, we conclude that the Gaussian Potts

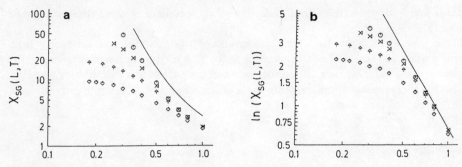

Figure 2:
a) Log-log plot of $\chi_{SG}(L,T)$ vs. $T$. Symbols have the same meaning as in Fig. 1. The full curve is a high-temperature expansion (which has two nontrivial terms only).
b) Log-log plot of $\ln \chi_{SG}(L,T)$ vs. $T$. In this plot eqs.(10,11) become a straight line with slope of minus $\sigma$, if the proportionally constant is put to unity. The straight line shown on the figure has slope $-2$, corresponding to the theoretical value $\sigma = 2$.

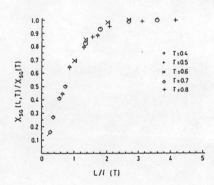

Figure 3: Scaling plot of $\chi_{SG}(L,T)/\chi_{SG}(T)$ against the scaled variable $L(T)/\ell(T)$. Here $\chi_{SG}(T)$ has been adjusted to get the fit and the characteristic lengths $\ell(T)$ are fit parameters.

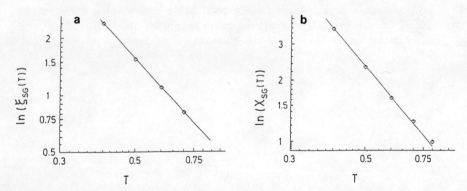

Figure 4:
a) Log-log plot of $\ln(\ell(T))$ vs. $T$. The straight line has slope $-1.97$, close to the value of $-2$ expected for a system at its lower critical dimension.
b) Log-log plot of $\ln(\chi(T))$ as a function of $T$. Here the slope of the full curve is $-1.95$, again very close to $-2$.

glass has a highly degenerate ground state due to the presence of the antiferromagnetic couplings /5/.

To conclude, we have shown that the 3 dimensional Potts glass behaves differently from Ising-spin glasses. Whereas the latter have a finite $T_c$ with power law divergencies, the Potts glass is right at its lower critical dimension and the correlation length diverges exponentially. Furthermore we found evidence for a macroscopically degenerate ground state, even for the short range model.

Acknowledgements:

One of us (M. S.) is supported by the Deutsche Forschungsgemeinschaft, SFB 262. The computations were performed on the CRAY-YMP 8/832 of the Höchstleistungs-Rechenzentrum (HLRZ) Jülich and on the VP100 of the Rechenzentrum Kaiserslautern. The work of A.P.Y. is supported in part by the National Science Foundation under grant DMR 87-21673.

# References

[1] D.J. Gross, I. Kanter and H. Sompolinsky, Phys. Rev. Lett. 55, 304 (1985)

[2] K. Binder and A.P. Young, Rev. Mod. Phys. 58, 801 (1986)

[3] R.N. Bhatt and A.P. Young, Phys. Rev. Lett. 54, 924 (1985);
Phys. Rev. B37, 5606 (1988)
and in *Proc. Heidelberg Colloquium on Glassy Dynamics*
(L. van Hemmen and I. Morgenstern, eds.), p.215, (Springer, Berlin 1987)

[4] W.L. McMillan, J. Phys. C17, 3179 (1984)

[5] Further evidence for a highly degenerate ground state comes from a numerical transfer-matrix calculation on 2 dimensional strips where we obtained a finite zero temperature entropy for the Gaussian Potts glass (to be published).

# Numerical Transfer Matrix Studies of Ising Models

M.A. Novotny

Supercomputer Computations Research Institute, Florida State University,
Tallahassee, FL 32306, USA

**Abstract.** Numerical transfer matrix methods are introduced which give scaling results for Ising models in high dimensions ($d \leq 7$) and for dimensions between 1 and 2. Results for both the critical temperature and the thermal critical exponent are presented.

## 1. Introduction

Two powerful methods of investigating the critical behavior of Ising models are the use of Monte Carlo simulations and the use of transfer matrix studies. Whereas Monte Carlo simulations have been applied to high dimensional models,[1] transfer matrix methods have thus far been useful only for two dimensional systems. However, some transfer matrix results have been obtained in three dimensions.[2] Thus far both numerical methods have only been applicable to systems in integer numbers of dimensions, although some previous numerical investigations have tried unsuccessfully to calculate quantities in non-integral dimensions.[3,4] In this short paper results for transfer matrix studies of Ising models in integer dimensions $d = 3, 4, 5, 6, 7$ and for $1 < d \leq 2$ will be presented. Because of the constraint on the length of the paper, only a brief sketch of the methods and some results are presented here. Future publications will describe the methods in more detail and present more complete results.

## 2. Models and Methods

The Ising model is studied on a hypercubic lattice, with a Hamiltonian given by $\mathcal{H} = -J\sum_{nn} s_i s_j$, where $s = \pm 1$ and the summation is over the $2d$ nearest-neighbor bonds. Helical boundary conditions within the $d-1$ dimensional hypercube of the transfer matrix are chosen, while periodic boundary conditions in the transfer direction are used. In the standard fashion the partition function for the model with $N$ spins in the $d-1$ hypercube and $M$ such hypercubes is given by

$$Z = \mathrm{Tr}\left(\left(\mathbf{A}\mathbf{D}_{d-1}\right)^M\right) \qquad (1)$$

Here $\mathbf{A}$ is a $2^N \times 2^N$ matrix formed from the direct product of $N$ $2\times 2$ two-body transfer matrices $\mathbf{a}$ with elements $\langle s_i|\mathbf{a}|s_j\rangle = \exp(Js_is_j/k_BT)$, where $k_B$ is Boltzmann's constant and $T$ is the temperature. For $d \geq 2$ the matrix $\mathbf{D}_{d-1}$ is given by[3,5]

$$\mathbf{D}_{d-1} = \mathbf{I} \odot \left(\mathbf{PA}\right) \odot \left(\mathbf{P}^{N^{\frac{1}{d-1}}}\mathbf{A}\right) \odot \left(\mathbf{P}^{N^{\frac{2}{d-1}}}\mathbf{A}\right) \odot \cdots \odot \left(\mathbf{P}^{N^{\frac{d-2}{d-1}}}\mathbf{A}\right). \quad (2)$$

Here $\mathbf{I}$ is the identity matrix, $\odot$ refers to Hadamard (element by element) matrix multiplication and the permutation matrix $\mathbf{P}$ is given and described in ref. 3. Although scaling can be performed with the transfer matrices given by Eq. 2,[6] the interactions obtained by this procedure are not symmetric if $N^{\frac{1}{d-1}}$ is not an integer. However, it is also possible to symmetrize the interactions. For $1 \leq d \leq 2$ the symmetrized diagonal matrix which interpolates between $d=1$ and $d=2$ as $\eta$ is varied between 0 and 1 is given by

$$\mathbf{D}_\eta = \mathbf{I} \odot \left(\mathbf{P}^{N^\eta}\bar{\mathbf{A}}\right) \odot \left(\mathbf{P}^{N^{-\eta}}\bar{\mathbf{A}}\right). \quad (3)$$

Here $\bar{\mathbf{A}}$ is the direct product matrix described above, but with an interaction strength $J/2$. A transfer matrix study of the unsymmetrized version of this model was presented in ref 3. It is also possible, both for $1+\eta$ and for high dimensions, to perform an interpolation in the Hamiltonian as described in ref. 4. In this case the symmetrization is automatic.

In the $M \to \infty$ limit the correlation length is given by $\xi^{-1} = \ln|\lambda_1/\lambda_2|$ with $\lambda_i$ the $i^{th}$ largest eigenvalue of the matrix $\mathbf{AD}_{d-1}$. The scaling form for $\xi$ is[7]

$$\xi(N) = N^{\frac{\omega}{d-1}} R\left(tN^{\frac{\Theta}{d-1}}\right). \quad (4)$$

This gives the phenomenological scaling criterion for $T_c$ to be[8,9]

$$\frac{\xi(N)}{N^{\frac{\omega}{d-1}}} = \frac{\xi(N')}{N'^{\frac{\omega}{d-1}}}. \quad (5)$$

The thermal exponent relation

$$\Theta + \omega = \left[(d-1)\ln\frac{d\xi(N)/dT}{d\xi(N')/dT}\right] \times \left[\ln\frac{N}{N'}\right]^{-1}. \quad (6)$$

was also calculated. For systems which satisfy hyperscaling $\omega = 1$ and $\Theta = y_T = \nu^{-1}$. However, for Ising models in $d > 4$ hyperscaling is not obeyed[1,10] because of the existence of a dangerous irrelevant scaling variable.[11,12] In this case the scaling depends on the boundary conditions. It has been postulated that for periodic, and probably helical, boundary conditions $\omega = (d-1)/3$ and $\Theta = \frac{2}{3}(d-1)$.[7]

## 3. Data, Analysis, and Discussion

Results will be presented for interpolation using the transfer matrix for both the asymmetric case (TM-A) and the symmetric case (TM-S) which were briefly described above. Results will also be presented for interpolation in the Hamiltonian (H).

Table 1 gives the thermal exponent relation $\Theta+\omega-1$ from Eq. 6 for the Ising model for $3 \leq d \leq 7$. Table 2 gives the location of the critical temperature, $T_c$. The value $\omega=1$ was assumed for $d \leq 4$ while $\omega=(d-1)/3$ was assumed for $d \geq 4$. The results presented here are for strips with $N=11$ and $N'=13$. The exponent relations for both the TM-S and the H cases are in very reasonable agreement with the expected values. It is not surprising that the values for

**Table 1** The value of $\Theta+\omega-1$, given by Eq. 6, is shown for the Ising model in 3 through 7 dimensions. For systems which satisfy hyperscaling this relation gives $y_T = 1/\nu$. The scaling was performed using strips with 11 and 13 spins. The interpolation is performed either asymmetrically in the transfer matrix (TM-A), symmetrically in the transfer matrix (TM-S), or in the Hamiltonian (H).

| $\Theta+\omega-1$ for $d$ dimensional Ising Model | | | | |
|---|---|---|---|---|
| $d$ | TM-A | TM-S | H | Expected |
| 3 | 2.84 | 1.70 | 1.62 | 1.59 |
| 4 | 2.98 | 2.18 | 1.99 | 2.00 |
| 5 | 3.14 | 3.02 | 3.07 | 3.00 |
| 6 | 4.55 | 4.03 | 3.79 | 4.00 |
| 7 | 4.89 | 4.58 | 4.55 | 5.00 |

**Table 2** The critical temperature $T_c$, given by Eq. 5, is shown for the Ising model in 3 through 7 dimensions. The scaling was performed using strips with 11 and 13 spins. The interpolation is performed either asymmetrically in the transfer matrix (TM-A), symmetrically in the transfer matrix (TM-S), or in the Hamiltonian (H). The expected values for dimension $d$ are from the references listed; ($d$: ref.) (3: 13,14) (4: 15,16) (5: 1) (6: 17) (7: 17).

| $k_B T_c/J$ for $d$ dimensional Ising Model | | | | |
|---|---|---|---|---|
| $d$ | TM-A | TM-S | H | Expected |
| 3 | 6.79 | 5.23 | 4.80 | 4.51 |
| 4 | 9.21 | 7.99 | 7.43 | 6.66 |
| 5 | 8.51 | 8.09 | 7.90 | 8.77 |
| 6 | 14.01 | 13.06 | 12.50 | 10.84 |
| 7 | 12.69 | 12.01 | 11.90 | 12.87 |

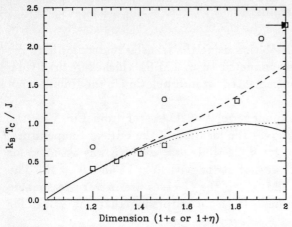

**Fig. 1** The critical temperature, $k_B T_c/J$ is shown for two interpolation schemes between 1 and 2 dimensions. The squares are for a symmetric interpolation in the transfer matrix (TM-S) while the circles are for interpolation in the Hamiltonian (H). The strip widths used are $N=9$ and $N'=11$. Also shown are $1+\epsilon$ expansions to two-, three-, and four-loop order corresponding to dotted, dashed, and solid lines respectively from ref. 18, 19, and 20. The arrow shows the exact result in $d=2$.

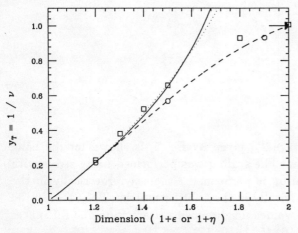

**Fig. 2** The thermal exponent $y_T$ is shown for two interpolation schemes between 1 and 2 dimensions. The legend is the same as for Fig. 1.

the TM-A case are not very good, since the effective Hamiltonian[3] in this case is not symmetric. Consequently, it is important to symmetrize the diagonal matrix in the transfer matrix to obtain reasonable results. One word of caution should be injected here. While in normal cases the phenomenological scaling produces values which monotonically converge to the exact value as $N$ and $N'$ are increased, that is not true for either the TM-A[6] case or the other two cases.

The results for interpolation between dimensions 1 and 2 are presented for $T_c$ in Fig. 1 and for the exponent $\Theta = y_T = 1/\nu$ in Fig. 2. In these figures the assumption that $1+\eta$ gives the dimension was made. It would, however, be better to obtain the dimension by using the scaling of Eq. 4. This will be done in a future publication. For comparison, results from the $1+\epsilon$ expansions from ref. 18, 19 and 20 are shown, even though some arguments have been presented[21] that these results may not represent the Ising model. From the two figures it is seen that both the TM-S and H interpolation schemes give results which are anticipated for the Ising model in dimensions between 1 and 2. In particular, both $T_c$ and $y_T$ seem to monotonically increase with dimension and approach the exact results at $d=1$ and $d=2$.

**Acknowledgements.** This research is supported in part by the Florida State University Supercomputer Computations Research Institute, which is partially funded through Contract # DE-FC05-85ER25000 by the U.S. Department of Energy. Useful discussions with M. Grant, V. Privman, P. A. Rikvold, J. Viñals, and R. K. P. Zia are acknowledged.

### References

[ 1] K. Binder, *Z. Phys. B* **61**, 13 (1985).
[ 2] M. Henkel, *J. Phys. A* **20**, 3969 (1987).
[ 3] M. A. Novotny, in *Quantum Monte Carlo Methods in Equilibrium and Nonequilibrium Systems*, ed. M. Suzuki, Springer Series in Solid-State Sciences, Volume 74, (Springer-Verlag,Berlin,1987).
[ 4] M. A. Novotny, *J. Appl. Phys.* **63**, 3546 (1988).
[ 5] M. A. Novotny, *J. Math. Phys.* **20** 1146, (1979); *ibid* **29** 2280 (1988).
[ 6] M. A. Novotny, *J. Appl. Phys.* (in press).
[ 7] V. Privman in *Finite Size Scaling and Numerical Simulation of Statistical Systems*, ed. V. Privman (World Scientific, in press).
[ 8] M. P. Nightingale, *Physica A* **83**, 561 (1976).
[ 9] M. P. Nightingale, *Phys. Lett. A* **59**, 486 (1977).
[10] K. Binder, M. Nauenberg, V. Privman and A. P. Young, *Phys. Rev. B* **31**, 1498, (1985).
[11] V. Privman and M. E. Fisher, *J. Phys. A* **16**, L295, (1983).
[12] V. Privman and M. E. Fisher, *Phys. Rev. B* **30**, 322, (1984).
[13] G. S. Pawley, R. H. Swendsen, D. J. Wallace, and K. G. Wilson, *Phys. Rev. B* **29**, 4030, (1984).
[14] P.-Y. Lai and K. K. Mon, *Phys. Rev. Lett.* **62**, 2608, (1989).
[15] H. W. J. Blöte and R. H. Swendsen, *Phys. Rev. B* **22**, 4481, (1980).
[16] O. G. Mouritsen and S. J. Knak Jensen, *Phys. Rev. B* **19**, 3663, (1979).

[17] M. E. Fisher and D. S. Gaunt, *Phys. Rev.* **133A** 224 (1964).
[18] D. Wallace and R. K. P. Zia, *Phys. Rev. Lett.* **43** 808 (1979).
[19] D. Forster and A. Gabriunas, *Phys. Rev. A* **23** 2627 (1981).
[20] D. Forster and A. Gabriunas, *Phys. Rev. A* **24** 598 (1981).
[21] D. A. Huse, W. van Saarloos, and J. D. Weeks,
*Phys. Rev. B* **32** 233 (1985).

# A Computer Simulation of Polymers with Gaussian Couplings: Modeling Protein Folding?

*H.-O. Carmesin and D.P. Landau*

Center for Simulational Physics, University of Georgia, Athens, GA 30602, USA

## 1. Introduction

A protein is a chain molecule which is characterized chemically by the sequence of amino acids (monomers) which form it. At sufficiently high temperature T, the protein exhibits a rather random and loose conformation, and as T is lowered it undergoes a "protein folding transition" [1,2] to one of a few rather tight and specifically ordered conformations. In an organism, a protein occurs in a folded state, the specific nature of which determines its "protein functions". Perhaps the most important question currently facing microbiologists is the nature and origin of protein folding.

While there exists a variety of proteins (substructures of which can be identified and classified [3] and are sometimes disordered [4]), it is possible to identify within certain ensembles of proteins scientifically meaningful correlations [5–7] as well as evolutionary rules [1]. Therefore, statistically simple ensembles of heteropolymers have been suggested as models for ensembles of proteins [8–10]. These are related to Ising spin glasses, Potts glasses [11] and multipolar glasses [12], all of which have been studied extensively.

Since the time scales involved in protein folding range at least from 100 picoseconds to 100 seconds [1,13], a computer simulation of the entire process is not currently available [1]. However, simulations of certain aspects of protein folding are possible [1,14]; in particular sophisticated Monte Carlo [15,16] and Langevin [17] dynamics enable one to stochastically simulate motions which take place at very different time scales.

## 2. Model

The excluded volume of solvent–particles and monomers is modelled by a simple Yukawa potential, since we are interested in the effect of random coupling among monomers rather than in steric or potential–shape effects. Monomers of chemical distance one (*i.e.*, directly chemically connected) are bound by an exponential function. We model the hydrogen bond coupling between monomers (the strength of which depends on the specific amino acids and their environment) with random bonds [8–10] of Gaussian distribution between each two monomers. The total potential energy is thus

$$\varphi = \epsilon \sum_{<i,j>} e^{-r_{ij}}/r_{ij} + 5\epsilon \sum_{<i,j>_c} e^{r_{ij}} + \sum_{<i,j>_m} e^{-r_{ij}} J_{ij}, \qquad (1)$$

where $<i,j>$ denotes all pairs of particles, $<i,j>_c$ all pairs of particles with chemical distance one, $<i,j>_m$ all pairs of monomers within the chain, and $r_{ij}$ the distance between particles i and j. The coupling constants $J_{ij}$ are distributed via

$$P(J_{ij}) = \frac{1}{\sqrt{2\pi}\, J\epsilon} \exp\left[\frac{(J_{ij}+J\epsilon)^2}{2(J\epsilon)^2}\right]. \qquad (2)$$

The mass of each particle is chosen to be $26.66*10^{-27}$ kg, $\epsilon = 1.67*10^{-21}$ Joule, the unit of length is 3Å, we choose $J = 40$ (or 12) and the T–unit = 120 000K.

## 3. Simulation

The system is placed in a cubic box with periodic boundary conditions (pbc), while the coupling via J is evaluated without pbc. We model 81 particles, 44 of which are monomers forming a chain molecule, while 37 represent the solvent. (We investigate the effect of finite chain length L by simulating systems containing either two chains with L = 22 or four chains with L = 11. We simulate Langevin dynamics (at each step all velocities are rescaled by the friction factor $f_0$, the stochastic force is chosen appropriately) and with a vectorized algorithm obtain 10 steps per second on an IBM3090.

We prepare an unfolded state for the simulation of folding and a folded state for the simulation of unfolding as follows. Initially the conformation is a Gaussian random walk and the solvent–particles are distributed uniformly. T is increased (exponentially fast, so that the velocities exhibit reasonable directions) from nearly zero to 0.3, where it stays for $t_0/3$ ($t_0$ = "preparation time") and equilibrates rapidly. Then T is lowered to approximately 1.5 (0.5) $T_{fold}$ for the simulation of folding (unfolding), where the respective chains fold at $T = T_{fold}$. At $t = 2t_0/3$, T is adjusted to its desired value and does not change systematically until $t_0 + t_{obs}$ ($t_{obs}$ observation time during which data are sampled).

## 4. Results for the solvent

At sufficiently low T, our simulation of the classical equations of motion becomes inadequate since the uncertainty principle is violated and quantum fluctuations become relevant. We estimate the line $T_{uc}(\rho)$, at which the action of a solvent atom $\Delta x \Delta p = h/2\pi$ ($\Delta x$ is the mean distance a solvent–particle moves before returning, $\Delta p$ its mean momentum), is shown in Fig. 1. The sharp increase of $T_{uc}(\rho)$ at low T is caused by the fact that particles repel each other rather exponentially, while at high T, the Coulomb part of the Yukawa potential dominates. The increase at low density ($\rho < 100$), is well fitted, Fig. 1, by

$$T_{uc} = \text{const.} \, (\rho - \rho_{fictive})^\alpha. \qquad (3)$$

$\alpha \approx 5.5$, and $\rho_{fictive} \approx 204$ is the density at which the pressure diverges for a potential that is identical to this Yukawa potential at large interparticle–distance [18]. $\rho_{fictive}$ is also relevant for solidification [18] and thus for the stability of the proteins: The heat capacity of folding is nearly constant for $T > T_{solid}$ (solidification temperature of solvent) [1,19,20]. This implies that there is a temperature $T_{max} < T_{fold}$ at which the folded state exhibits maximum thermodynamic stability [20]. There are proteins in nature for each of the three resulting possibilities, $T_{solid} < T_{max}$, even low T denaturation has been observed [20], $T_{max} < T_{solid} < T_{fold}$ [20], low temperature unfolding is excluded, $T_{fold} < T_{solid}$ [21] "slaved

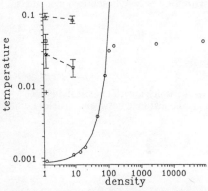

Figure 1: Boundaries in the $\rho$–T–plane: Data for limits of validity for the classical model $T_{uc}$ (O); data for $\rho$–dependence of the folding transition $T_{fold}$ (L = 44, J = 40, $\Delta$), (L = 44, J = 12, *), (L = 22, J = 40, □), (L = 11, J = 40, +). The solid curve comes from Eq. 3.

## 5. Results for protein folding

We find indications of protein folding in the change in the mean squared radius R ($R = \sqrt{\Sigma(r_i-r_o)^2}$ where $r_o$ is the location of one end of the chain), see Fig. 2, and the internal energy [22], and observe hysteresis in both quantities, in qualitative agreement with slow relaxation in proteins [1]. $T_{fold}$ decreases with $\rho$, Fig. 1, in qualitative agreement with experiments [23]. $T_{fold}$ increases with L, Fig. 1, but this size dependence shrinks rapidly for increasing L. {Since the interaction is of short range [1], bulk and surface are distinguished sharply. Consider for simplicity complete folding into a cubic shape. One finds for $L = 27$ $n_{int} = 1$ (particles in interior) and $n_{surface} = 26$ (particles at surface). For $L = 1000$ $n_{int} = 512$ and $n_{surface} = 488$. Thus, the internal energy per particle cannot be decreased significantly by increasing L much more than 1000.} Thus, 1000 could be seen as a reasonable maximum substructure size, while a minimum substructure size seems to be $L \approx 11$, for which $T_{fold} = 0$ cannot be excluded, see Fig. 1, both in approximate agreement with experiment [3].

Conformational properties have been analyzed in previous simulations by various ad hoc approaches [1], while a systematic investigation of all order parameters is provided by the mapping to multipolar glasses [10] (which applies to any distribution of $J_{ij}$ and especially to realistic proteins). A systematic investigation of the total overlap is sufficient to identify folding into preferential conformations. We define (in reasonable approximation of $\delta$-functions) the total indexed overlap [10] of the ith monomers in two different conformations (coordinates $r_i$ and $s_i$)

$$Q_i(r_i,s_i) = \exp(5|r_i-s_i|) - Q_i^0(r_i,s_i) , \qquad (4)$$

where $Q_i^0(r_i,s_i)$ is the respective overlap in the uncoupled ($J = 0$) system. The total overlap is

$$Q(r_i,s_i) = 1/L \sum_i^L Q_i(r_i,s_i) . \qquad (5)$$

For a given bond configuration, the autocorrelation function then reads

$$q_{autocorr}(t) = Q(r_i(t_o),r_i(t_o+t)) . \qquad (6)$$

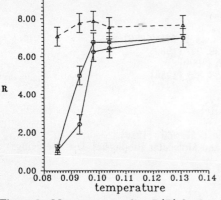

Figure 2: Mean square radius R(T) for $L = 44$, $J = 40$, folding (□) and unfolding (○), as well as for $L = 44$, $J = 0$ (△). The system has been simulated with $t_o = 3500$, $t_{obs} = 1500$ and friction factor $f_o = 0.45$.

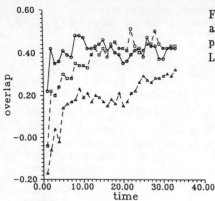

Figure 3: The functions $q_{autocorr}$ (○), $q_s$ (□) and $q_u$ (△) are shown. The simulation is characterized by the parameters $t_o = 500$, $t_{obs} = 6600$, $T = 0.015$, $J = 12$, $L = 44$, $f_o = 0.45$, the unit of time is 200 steps.

We also define $q_s(t)$ as the equivalent of Eq. 6 for the same bond configuration and another random initial configuration, and $q_u(t)$ as the equivalent of Eq. 6 for a different bond configuration and another random initial configuration.

From Fig. 3 we conclude that chain molecules with identical couplings gain maximum possible overlap after 3000 steps, while chains with uncorrelated couplings increase their mutual overlap so slowly that our simulation is too short to identify its equilibrium value.

## 6. Conclusion

We find that a random Gaussian heteropolymer folds into specific preferred conformations at $T_{fold}$, and exhibits hysteresis while unfolding. The stability of substructures of different size is discussed. The overlap between identical heteropolymers relaxes faster than that between different ones, and a more systematic investigation of the dynamics and stability is in preparation [22]. Solvent–particles which are simply repulsive hinder folding increasingly at higher $\rho$. At sufficiently large $\rho(T)$ they solidify and thereby hinder folding completely.

## References

1. C. Gheles and J. Yon, **Protein Folding**, Academic Press, New York (1982).
2. R. Jaenicke, **Protein Folding**, North–Holland, Amsterdam (1984).
3. J.S. Richardson, in Ref. 2.
4. R.L. Baldwin and T.E. Creighton, in Ref. 2.
5. F.M. Pohl, in Ref. 2.
6. M.F. Perutz in Ref. 2.
7. G.E. Schulz, in Ref. 2.
8. E.J. Shakhnovich and A.M. Gutin, Europhys. Lett. **8**, 327 (1989).
9. H.–O. Carmesin and T. Vilgis, Phys. Lett **A138**, 227 (1989).
10. H.–O. Carmesin, to be published in J. Phys. A (1990).
11. K. Binder and A.P. Young, Rev. Mod. Phys. **58**, 821 (1986).
12. H.–O. Carmesin, Z. Phys. B**73**, 381 (1988).
13. R.L. Baldwin in Ref. 2.
14. D.E. Stewart. A.Sarkar and J. Wampler, preprint.
15. R.H. Swendsen, in these proceedings.
16. R.H. Swendsen and J.–S. Wang, Phys. Rev. Lett. **57**, 2607 (1986).
17. T. Schlick et al., in **Theoretical Biochemistry and Molecular Biophysics**, D.L. Beveridge and R. Lavery eds., in press.
18. H.–O. Carmesin, Phys. Rev. B**41**, 4349 (1990).
19. J. Schellman, Biopolymers **17**, 1305 (1978).
20. J. Schellman, in Ref. 2.
21. I.E.T. Iben et al., Phys. Rev. Lett. **62**, 1916 (1989).
22. H.–O. Carmesin and D.P. Landau, in preparation.
23. R. Jaenicke and R. Rudolph, in Ref. 2.

# Numerical Studies of Absorption of Water in Polymers

*J.L. Vallés*[1,2], *J.W. Halley*[1], *and B. Johnson*[1]

[1]School of Physics and Astronomy, University of Minnesota,
Minneapolis, MN 55455, USA
[2]Departament de Fisica Fonamental, Universitat de Barcelona,
Avgda. Diagonal 647, 08028 Barcelona, Spain

**Abstract.** We present two computer simulation studies on water penetration into polymers. A molecular dynamics model for the interaction between water molecules and a polymer of industrial interest allows us to determine where the water is bound and the shifts that this produces in the infrared spectrum. As a limiting case for diffusion of water in glassy polymers, we describe also Monte Carlo simulations for diffusion with trapping on a percolation lattice.

In many technological applications of polymers the penetration of water can influence negatively the desired physical and/or chemical properties. The presence of water inside polymers and the structure of the water-polymer system have been the subject of research for a long time [1], although most work concentrates on problems of biological interest [2]. We have carried out simulation studies of the absorption of water in polymers that are focused on two different levels. In the more microscopic level we have performed a molecular dynamics simulation [3] of the structure and dynamics of a water-polymer model and analyzed the infrared absorption by $D_2O$ molecules in interaction with a monomer characteristic of several polymer adhesives. On a more macroscopic level we have studied [4] via Monte Carlo the diffusion with interactions and trapping on realizations of the two dimensional percolation model. We plan to compare all our numerical results with experiments in progress in the IBM Almaden Research Center.

A three-dimensional view of the water and the polymer repeat unit used in our molecular dynamics simulation is shown in Fig. 1. In order to construct our system we started from a set of 216 water molecules that had reached equilibrium at room temperature. The numerical descrip-

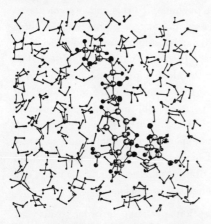

Fig. 1. Side view of the three-dimensional water-polymer system after equilibration.

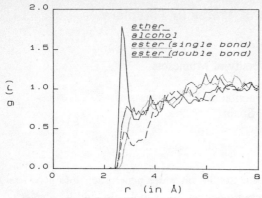

Fig. 2. Radial distribution function of the oxygens in water molecules around oxygens in the different polymer groups.

tion of the monomer, obtained with a molecular mechanics program [5], was then embedded by superimposing the positions of the 71 monomer atoms and removing the water molecules that were closer than $2.15\text{Å}$ to any non-hydrogen polymer atom. We were thus left with 192 water molecules in the sample, which were allowed to move in accordance with Newton's equations. The simulation was microcanonical, and was done with periodic boundary conditions, a water density of $1g/cm^3$ and a temperature of $300K$. The monomer atoms were kept in their original positions.

Our water molecules were represented by a central force molecular dynamics model [6] which takes into account the intramolecular dynamics. To describe the intermolecular interactions each oxygen and hydrogen in the water is given a partial charge, this requiring the use of Ewald summation techniques, and one considers also short range Lennard-Jones forces between the oxygens. The intramolecular forces originate from either a Morse potential or a harmonic potential between the hydrogens and the oxygen and a harmonic potential between the hydrogens. We computed the Coulomb and Lennard-Jones interactions between the polymer and the water molecules using the pair-potential functions suggested by Jorgensen and Tirado-Rives [7].

Our first concern was to find out whether the water molecules were in any way bound to the oxygen in the alcohol, ester and ether groups on the polymer. To this end, we calculated the radial distribution function of oxygen atoms in water around oxygen atoms in those three groups. Our time averages were over 120000 molecular dynamics steps corresponding to 48 ps of real time, after an equilibration of at least 15000 steps (6 ps). The results, presented in Fig. 2, are quite conclusive: the sharp peak obtained for the alcohol group and the absence of such a peak for the oxygen in the other two groups show that here the water molecules bind only at the alcohol group. This is in agreement with experiments [8] unknown to us prior to our simulation indicating that water binds to hydroxyl but not to ester groups in branched, oxidized polyethylene. Due to the symmetry of the monomer, we can study independently the time averaged contributions of chemically equivalent oxygens with different geometry and environment. In an evolution as long as ours these contributions are essentially identical, and this tells us that letting the polymer move would not change the conclusions.

Next, we calculated in our model the infrared absorption spectrum, in order to discover whether the water bound at the alcohol group could be distinguished from the rest of the water

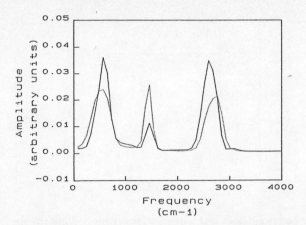

Fig. 3. Fourier transform of the dipole-dipole autocorrelation function in the $D_2O$ molecules in the case of bound water (solid line) and unbound water (dotted line), as explained in the text.

molecules by a shift in the positions of the absorption peaks. In these calculations we simulated heavy water ($D_2O$), because the monitoring of the $OD$ peak in polymers exposed to heavy water is a useful experimental technique to study the diffusion of water into polymers. In Fig. 3 we show the Fourier transform of the dipole-dipole autocorrelation function for the unbound bulk heavy water, together with the one for the bound heavy water. This latter curve is obtained by substracting from the signal calculated from the $D_2O$ molecules whose oxygen atom is within 3.3 Å of an alcohol of the polymer a background corresponding to the bulk $D_2O$. The stretch, scissors and librational peaks appear clearly in both of the curves. We can conclude that the librational peak of the bound water spectrum is shifted up by about $40 cm^{-1}$ while the stretch peak of the same spectrum is shifted down by about $70 cm^{-1}$. These shifts are large enough to be observed experimentally.

The diffusion of water molecules into a glassy polymer can also be described with a simple extension of the model of a particle diffusing on a percolation cluster. In the new version, one has to take into account that the penetrating particles will be trapped sometimes at polymer sites, and that the density of those particles may be too high to ignore the interactions between them. This is also a many-particle version of the dual-mode transport model that was studied by Fredrikson and Helfand [9], where one considers two types of water which are interpreted as absorbed and dissolved water. In the lattice model that we adopted, a fraction of the sites are picked randomly as black ("polymer") sites and water particles coming in from one of the walls hop from site to site. Double occupancy of any site is forbidden and the flux at the wall is controlled by a chemical potential. We have performed computer simulations on this lattice in two dimensions, in the case when the concentration of white ("non-polymer") sites is near the percolation concentration. The evolution is governed by the following rates: we accept all attempts from a particle in a white site going to a white or to a black site, but particles are permanently trapped on a black site once they land there. As a result, the particles move through a region of the lattice with filled trapping sites towards the front, where they are trapped. One finds two different behaviors for the time evolution of the density distribution of trapped and untrapped water, as it is shown in Fig. 4. This diffusion process may also be described by a

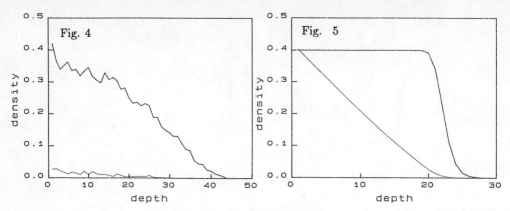

Fig. 4. Density distribution of trapped (solid line) and untrapped water (dotted line) at the end of one of our Monte Carlo runs.

Fig. 5. Density distribution of trapped (solid line) and untrapped water (dotted line) obtained from the mean-field equations in a typical case.

mean field model [4], and the numerical solutions exhibit fronts (Fig. 5) qualitatively similar to the ones found in the simulation, thus suggesting that the occurrence of fronts is associated with the nonlinearities and not with the percolation character of the lattice model.

In our simulations the lattice has a width of 1000 sites and a depth of 100 sites. Some details of the optimization of the algorithm have been described elsewhere [4]. Our evolution times extend up to 500 "steps" per surface spin and we averaged over 10 initial configurations. When we study the first moment $R$ of the density profile of the trapped water versus simulation time, we find clear evidence of deviation from normal diffusion. Assuming that the geometrical properties of the homogeneous medium are those of a realization of the percolation model, we can try [10] a scaling form:

$$R = t^{1/2} |\Delta p|^{\mu'/2} F(|\Delta p|^{\mu'} t^{x\mu'}) . \qquad (1)$$

Here $t$ is the Monte Carlo time and $\Delta p = p - p_c$, where $p$ is the concentration of white sites and $p_c = 0.5928$ is the percolation concentration. The exponent $\mu'$ is the conductivity exponent, and it is discussed below. Analyzing the asymptotic behavior one can derive an expression giving the value of $x$. In the case when $p < p_c$, we know that after a long time the penetration of the particles will have stopped, and the first moment will depend on $\Delta p$ as

$$R \propto |\Delta p|^{\beta/2-\nu}, \quad t \to \infty, \quad p < p_c \qquad (2)$$

where the exponents $\beta$ and $\nu$ are percolation exponents, in accordance with our assumption. When $p > p_c$, on the other hand, the first moment will increase at long times as

$$R \propto (Dt)^{1/2}, \quad t \to \infty, \quad p > p_c . \qquad (3)$$

The diffusion constant $D$ will be considered to vanish as a power law in $\Delta p$ as $p \to p_c$, but here the value of the exponent may be different from the one found in diffusion on a percolation cluster, because the trapping modifies the dynamics. Thus we assume $D \propto |\Delta p|^{\mu'}$, but we take $\mu'$ as not fixed at the value ($\approx 1.3$) found for diffusion on a percolation cluster. With these two

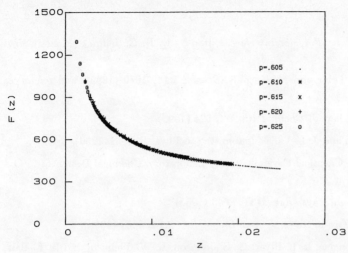

Fig. 6. Plot of $F(z) = R/(|\Delta p|^{\mu'/2} t^{1/2})$ versus $z = |\Delta p|^{\mu'} t^{x\mu'}$ for $\mu' = 1.5$, showing that scaling is satisfied.

constraints on the long time behavior, the scaling hypothesis is easily shown [10] to take the form (1), with

$$x = \frac{1}{2\nu + \mu' - \beta}. \quad (4)$$

We plotted our simulation data corresponding to several values of $p$ close to $p_c$ following (1), in order to test the scaling hypothesis for different values of $\mu'$. A good scaling is only obtained for $\mu = 1.5 \pm 0.1$, while for diffusion on a percolation cluster one has the value 1.3, which is outside this range. The best collapse is shown in Fig. 6.

We are presently carrying out simulations of diffusion in three dimensions in a system where, taking into account our findings in the molecular dynamics simulation, only about 5% of the black sites are traps for the water particles, and the preliminary data show good agreement with experimental results.

To summarize briefly, we have carried out numerical studies of the absorption of water in polymers at two different levels. In a molecular dynamics simulation of a water-polymer system we find that water absorbs at the oxygen in the alcohol groups but not at the oxygen in the ester or ether groups. The calculated infrared absorption spectrum shows a shift in the $OD$ stretch frequency at the bound water large enough so that the $D_2O$ methods could distinguish free from bound water. A Monte Carlo simulation of diffusion with trapping in a percolation lattice shows that the conductivity exponent is affected by the presence of trapping sites.

**Acknowledgements:** The work reported in this paper was supported in part by the Sponsored University Research program of IBM and by the Supercomputer Institute of the University of Minnesota.

# REFERENCES

1. *Water Structure at a Water-Polymer Interface*, edited by H. H. G. Jellinek (Plenum, New York, 1972).

2. J. Gao, K. Kuczer, B. Tidor and M. Karplus, *Science* **244**, 1070 (1989), and references therein.

3. J. L. Vallés and J. W. Halley, *J. Chem. Phys.* **92**, 694 (1990).

4. J. W. Halley, B. Johnson and J. L. Vallés, (submitted to Phys. Rev. B, 1990).

5. E. K. Davies, *Chemgraf*, Chemical Crystallography Laboratory, Chemical Design Ltd., Oxford.

6. K. Toukan and A. Rahman, *Phys. Rev.* **B31**, 2643 (1985).

7. W. L. Jorgensen and J. Tirado-Rives, *J. Am. Chem. Soc.* **110**, 1657 (1988).

8. D. W. McCall, D. C. Douglas, L. L. Blyer, G. E. Johnson, L. W. Jielinski, and H. E. Bair, *Macromolecules* **17**, 1644 (1984).

9. G. H. Fredrikson and E. Helfand, *Macromolecules* **18**, 2201 (1985).

10. D. Stauffer, *Introduction to Percolation Theory*, (Taylor and Francis, London and Philadelphia 1985), Chapter 5.

# Almost Markov Processes

*D. Bouzida*[1], *S. Kumar*[2], *and R.H. Swendsen*[1]

[1]Department of Physics, Carnegie Mellon University, Pittsburgh, PA 15213, USA
[2]Department of Physics, University of Pittsburgh, Pittsburgh, PA 15213, USA

A new approach to simulations of large molecules is described. By relaxing the usual restriction of algorithms to Markov processes, we are able to perform optimized simulations that take into account the inhomogeneity and anisotropy inherent in these systems. Our approach samples configurational space more efficiently than either standard Monte Carlo or molecular dynamics methods.

The methods described in this paper concern some work we have been doing recently on the simulation of large molecules: proteins, nucleic acids, etc. These simulations are becoming increasingly important in chemistry, biology, and medicine [1, 2]. Such calculations currently use an amount of supercomputer time comparable to that expended on lattice gauge theories, and it is growing constantly. Consequently, the efficiency of methods to simulate large molecules is of considerable importance.

There are many questions that biologists and chemists ask about the properties of large molecules. Some of these questions concern the short-time development of the molecule (i.e., a few hundred picoseconds). However, most of the questions concern equilibrium properties. The equilibrium configuration is of prime importance in determining biological function, and free-energy calculations are needed to predict reactions.

For questions regarding short-time dynamic behavior, it is of course necessary to do some form of molecular dynamics. However, for equilibrium properties we would expect Monte Carlo ($MC$) methods to be preferable. Interestingly enough, the chemical and biological communities express a strong preference for molecular dynamics methods in both cases [3]. This apparent inconsistency was one of the things that got us interested in the search for efficient methods for simulating large molecules.

The full story of what progress we have made in this search is too long to discuss in a brief talk. However, we would like to discuss our basic approach leaving some important features of how to deal effectively with the long-wavelength modes for another time.

Both Monte Carlo and molecular dynamics methods were originally developed to simulate fluids. It is clear from the structure of the methods that they exploit the fact that fluids are homogeneous and isotropic. For example, optimization of parameters like step sizes is normally done on a global basis. This is fine for a homogeneous system, but not for an inhomogeneous one; moreover, the structure of the anisotropy and inhomogeneity of macroscopic molecules changes with time.

This clearly points to the desirability of optimizing the parameters of the simulation locally, taking into account the development of the molecule as the simulation progresses. Unfortunately, if a Monte Carlo algorithm depends on anything other than the current configuration, the process is no longer Markovian.

We have lifted the restriction to Markov processes by making use of the information gathered during the simulation to optimize the local Monte Carlo moves [4]. We were then able to show by explicit simulations of models with exact solutions, that if reasonable limits are not exceeded, the resulting algorithms substantially increase the speed of the simulation without introducing any measurable deviation from the correct equilibrium behavior.

To account for the local anisotropy in the environment of each atom, MC trial moves are sampled from an ellipsoid, rather than the more usual sphere. Moves in an ellipsoid are generated by a transformation matrix $D_{ij}$, so that the trial move $d_i$ is generated by

$$d_i = \sum_j D_{ij} v_j \qquad (1)$$

where $v_j$ is a random vector chosen from a unit sphere. To calculate $D_{ij}$, we first solved the problem for an anisotropic simple harmonic oscillator ($SHO$) of the form

$$H = \frac{1}{2} \sum_{ij} k_{ij} x_i x_j \qquad (2)$$

where $k_{ij}$ is a real symmetric matrix. We can then show that $K_{ij} = \frac{1}{2}\beta k_{ij}$ can be found from the equation

$$[\beta \Delta E d_l d_m] = \sum_{ij} K_{ij} [d_i d_j d_l d_m] \qquad (3)$$

where $\Delta E$ denotes the change of energy for an attempted move $d_i$ and the square brackets indicate an average over all attempted moves, whether or not they were accepted. The matrix $D_{ij}$ is then given by

$$D_{in} = F(\lambda_n)^{-\frac{1}{2}} V_{in} \qquad (4)$$

where $\lambda_n$ is an eigenvalue of $K_{ij}$ and $V_{in}$ is the corresponding normalized eigenvector. F is a scale factor that is chosen to optimize the efficiency of the simulation as discussed below. $D_{ij}$ is then updated periodically to adapt to the changing local environment of each atom. Although this method, which we denote as Dynamically Optimized Monte Carlo ($DOMC$) is based on an analysis of a simple harmonic oscillator, these equations turn out to be remarkably robust when applied to a wide variety of Hamiltonians.

We have applied the method to a variety of systems. These include highly anisotropic one-, two-, and three-dimensional oscillators, systems of particles interacting through Lennard-Jones potentials, and biomolecules ranging from adenosine (one of the bases of $DNA$) to proteins with complicated structures such as the Bovine Pancreatic Trypsin Inhibitor ($BPTI$). In what follows, we will present results on simple models, and show explicitly how optimization is performed.

In one dimension, the quantities that enter into the optimization scheme are the maximum step size $\Delta$, the autocorrelation time $\tau$, the average acceptance ratio $<P>$, and both r.m.s displacements per MC-step defined as $\sqrt{<d_i^2>}$ and $<|d_i|>$, where the angle brackets indicate the thermal average. Different values of $\Delta$ optimize different quantities. For instance, for a $d = 1$ $SHO$, $V(x) = x^2$ with $\beta = \frac{1}{k_B T} = 1$, the maximum r.m.s displacement $\sqrt{<d_i^2>}$ is 0.6634 and corresponds to $\Delta = 2.62$, while $<|d_i|>$ has a maximum equal to 0.3746 corresponding to $\Delta = 1.75$. The correlation time has a minimum at $\Delta = 2.0$, with $\tau = 1.44$. These results show that optimization depends on what is being calculated. However it is important to note that the maxima and minima of these quantities are rather broad. Values of $F$ between 2 and 3 generally provide good optimization.

To test the systematic error, we simulated $V(x) = x^2$ with Eq. (3) evaluated over very short cycles. The exact value of the average energy $<E>$ is 0.5. Averaging over only 1 MC-step/cycle in Eq. (3), we find a large systematic error, with $<E> = 0.71$. However, with even 2 MC-steps/cycle, $<E> = 0.53$, and for 3MC-steps/cycle, $<E> = 0.506$. No systematic error was measurable for 4 MC-steps/cycle. Since we have generally used between 10 and 100 MC-steps/cycle in applications, the systematic error is completely negligible.

This method has also been applied successfully to $d = 2$ and $d = 3$ models with anisotropies in the coupling constants up to 1000. The acceptance ratio is independent of the anisotropy, but does depend on how the random jumps in the ellipse, or ellipsoid are generated. We have found that when they are generated uniformly in radius, the optimized acceptance ratio is ~50% for all dimensions, while it is ~40% for $d = 2$ systems and ~30% for $d = 3$ systems when the jumps are generated uniformly in area or volume.

As a first application to a molecule, we have performed $DOMC$ and molecular dynamics ($MD$) simulations on adenosine in the united-atom force field [5]. The energy-energy correlation time for the $DOMC$ simulations at 298K, including rotational moves, is 8.6 sweeps, and only 600 sweeps are necessary to reach the equilibrium r.m.s coordinate fluctuations (which are about 1.05Å). For an $MD$ simulation with a Verlet algorithm, this value is reached only after 80,000 MD-steps (80 picoseconds).

A simplified variation of $DOMC$, which is quite useful in some situations, is the Quasi-Optimized Monte Carlo ($QOMC$), in which the anisotropy is ignored, although the inhomogeneity is taken into account by optimizing the moves for each atom individually. Using the fact that the acceptance ratio is well approximated by a simple exponential function of the step size, the acceptance ratio is continuously monitored for each atom and the step size is reset periodically to optimize the simulation. This method can be particularly useful at high temperatures for simulated annealing.

Simulated annealing experiments on adenosine using $QOMC$ and rotational moves have been done, starting from random positions and lowering the temperature from $10^6 K$ to $1K$. We find that a total of 1000 sweeps is sufficient to reproduce the low-temperature configuration of adenosine.

# References

[1] C. H. Brooks III, M. Karplus and B. M. Pettitt, "Proteins: A theoretical perspective of dynamics, structure and thermodynamics", Advances in Chemical Physics, Vol. LXXI, edited by I. Prigogine and S. Rice, John Wiley and Sons, New York, 1988.

[2] J. A. McCammon and S. C. Harvey, "Dynamics of proteins and nucleic acids", Cambridge University Press, New York, 1987.

[3] S. H. Northrup, J. A. McCammon, Biopolymers **19** 1001 (1980)

[4] Since this workshop, in a talk at the APS March Meeting, S. Shumway and J. P. Sethna (Bull. Am. Phys. Soc. **35** 500 (1990)) have presented a novel technique for optimizing MC moves. Their specific technique is quite different from ours, but the essential spirit of their approach is very similar.

[5] S. J. Weiner, P. A. Kollman, D. A. Case, U. C. Singh, C. Ghio, G. Alagona, S. Profeta, and P. Weiner, J. Am. Chem. Soc. **106** 765 (1984)

# Vectorization of Diffusion of Lattice Gases Without Double Occupancy of Sites

*O. Paetzold*

Gesellschaft für Mathematik und Datenverarbeitung mbH
(German National Research Center for Computer Science)
P.O. Box 1520, W-5205 Sankt Augustin 1, Fed. Rep. of Germany

We present a new vectorizable algorithm for diffusion of lattice gases without double occupancy of sites. This algorithm is significantly faster than the algorithm usually used on scalar mainframes. Further we show the application of this algorithm to diffusion on percolation lattices at the percolation threshold.

## 1 Introduction

Diffusion in lattice gases is an important problem both from a fundamental and from an applied point of view. A physical example of lattice gases without double occupancy of sites and possible further interactions is the diffusion of atoms in interstitial and substitutional alloys. The collective and chemical diffusion has been studied by Monte-Carlo methods. For a review on the results of these investigations see reference [1]. The standard method used is described in [2]. To simulate hydrogen in metallic alloys modifications to the model have been made [3], but the algorithm used is still the standard one. The main disadvantage of this algorithm is that it is not vectorizable. So only scalar mainframes could be used. We will present a new fully vectorizable algorithm for lattice gas simulations, where double occupancy is excluded, which is significantly faster than the standard algorithm [4]. This algorithm has been applied to calculate the critical exponent of the root-mean-square displacement of tracer particles on percolation lattices at the percolation threshold.

## 2 The Standard MC-Algorithm

We will now describe the standard algorithm developed by Kehr et al. [2], which is an extension of an algorithm for the Kawasaki spin dynamics. Assume a lattice which is occupied by a number $N_p$ of particles. Because of the hard core interaction, only one particle can occupy a single site. During a time step $N_p$ random numbers between 1 and $N_p$ are chosen, two or more of which can be equal, allowing backward jumps of a particle to its previous site. The random numbers determine the 'name' of the particle. For a given particle a random jump direction is chosen randomly. A transition into this direction occurs only if this site is unoccupied.

## 3 The Vectorization Procedure

The main disadvantage of the above described procedure is that only one particle at a time is processed, although particles far apart would not influence each other if they tried to move simultaneously. The idea of vectorization is now to collect all those particles that can move independently of each other and process them simultaneously. Therefore, the particles must have a certain area around them into which other particles are not allowed to make a transition during a time slice. The minimum diameter of this area is three lattice constants in each lattice direction in the case of lattice gases with hard core interaction. So we subdivide the whole lattice into $3^d$ sublattices, where d is the spatial dimension. Instead of dividing a Monte-Carlo step into the number of particles on the lattice, the number of time slices is equal to the number of sublattices. During a time slice only one sublattice is processed as a whole, i.e., all particles on this sublattice try to perform a transition.

The sublattices are not processed in sequence but are randomly chosen, so that each sublattice can be processed several times in succession. Therefore, a particle which just has jumped onto a given sublattice can try a second jump during the subsequent time step, if that sublattice is chosen next. So backward jumps are still possible as in the standard algorithm.

A short remark on the number of sublattices is due. In two dimensions one can subdivide the lattice into five instead of nine independent sublattices [5]. In three dimensions there could be only seven instead of twenty-seven. But this subdivision becomes much more complicated, if further interactions are introduced. Therefore the application of the program to different problems and dimensions would require a reprogramming of parts of the algorithm. If one chooses the number of sublattices with respect to the linear distance of the interactions one has to change only a value in a parameter statement.

## 4 Verification and Comparisons

The vectorizable algorithm fulfills the condition of detailed balance. This can be seen easily. The transitions of particles on the same sublattice are independent and do not influence detailed balance. Therefore each particle on a sublattice is chosen randomly as the sublattice is. The probability of choosing a sublattice is constant. The same holds for the probability of a configuration, which is equal to $2^{-N}$. So the product of configuration probability and transition rate remains constant, which is the fulfillment of detailed balance.

To verify the correctness of the algorithm, we compared its results with analytical and numerical results. We found the correct behaviour of the mean-square displacement for all dimensions as well as the logarithmic correction in the tracer diffusion on a square lattice (cf. [6]).

We calculated the update rate for this algorithm and compared it with results of the standard method and with the update rate of a lattice gas without any interaction. At $c = 0.5$ on a square lattice, we found our update rate to be around 3.20 updates per microsecond on one processor of a CRAY X-MP/416, while the standard algorithm makes around 0.07 updates per microsecond on an IBM 3090-150. The lattice gas without interaction, which is nothing other than the 'ant in the labyrinth problem', has update rates of 4.6, see [7], and 3.4, see [8], updates per microseconds on a CDC Cyber 205 and CRAY Y-MP/832, respectively. The factor between our simulations and the much simpler case was expected to be much larger than two. This factor becomes much smaller if the concentration is increased. In the case described in the next section the algorithm slows down because of the low concentration and the factor between the two models is around 3.5.

# 5 Application to Diffusion on Percolation Lattices

We used the above mentioned algorithm to simulate lattice gases without double occupancy on a three-dimensional percolation lattice at the percolation threshold [9]. This was a recalculation of the work of Heupel [10]. Because of the restricted resources of the CDC Cyber 76 Heupel could only simulate lattices of size $20^3$ and very small particle concentrations. He calculated the critical exponent as

$$k = 0.20 \pm 0.01. \tag{1}$$

Our simulations[11] were done on a $30^3$ lattice, because we found, as Heupel did, that larger lattices would not increase the accuracy. But we were able to simulate lattice gases with much higher concentrations and found a new phenomenon. Namely, the exponent k becomes $0.26 \pm 0.01$ in a certain time range, see Figure 1, before it tends to an asymptotic value, which is equal at all concentrations. This crossover time depends on the concentration. A possible explanation for the plateau value is that the influence of the neighbouring particles is dominant during this time range. The percolation effect becomes dominant at later times.

In contrast to Heupel we evaluated the critical exponent as

$$k = 0.18 \pm 0.01, \tag{2}$$

which corresponds to the value of Roman [8].

Other applications of this algorithm are e.g. the diffusion of lattice gases through lattices with trapping sites, the diffusion of particles with nearest neighbour interaction and the interdiffusion of different particles. During the preparation of this manuscript we received a preprint where a similar algorithm is applied to the simulation of lattice polymers [12].

Thanks are due to K. W. Kehr, K. Kremer, D. Stauffer and R. Kutner for helpful discussions. The author thanks the Forschungszentrum Jülich GmbH (KFA) for their hospitality.

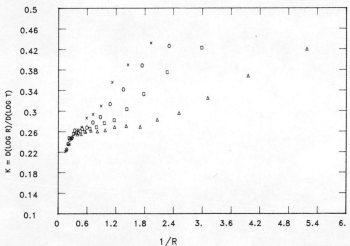

Figure 1: The exponent k versus the inverse root-mean-square displacement R for lattice gas with excluded volume interaction on a three-dimensional percolation lattice at the percolation threshold for the concentrations 0.3 (**X**), 0.5 (**O**), 0.7 (**◻**), 0.9 (**△**).

# References

[1] K. W. Kehr, K. Binder, in *Applications of the Monte Carlo Method in Statistical Physics*, ed. by K. Binder, Topics in Current Physics Vol. **36** (Springer, Berlin 1987), p. 181

[2] K. W. Kehr, R. Kutner, K. Binder, Phys. Rev. **B 23**, 4931 (1981)

[3] E. Salomons, J. Phys. C:Solid State Phys. **21**, 5953 (1988)

[4] O. Paetzold, submitted to Comp. Phys. Comm.

[5] R. B. Selinger, private communication

[6] R. Kutner, K. Binder, K. W. Kehr, Phys. Rev. **B 26**, 2967 (1982), and Phys. Rev. **B 28**,1846 (1983), H. van Beijeren, K. W. Kehr, R. Kutner, Phys. Rev. **B 28**, 5711 (1983), H. van Beijeren, R. Kutner, Phys. Rev. Lett. **55**, 238 (1985)

[7] R. B. Pandey, D. Stauffer, J. G. Zabolitzky, A. Margolina J. Stat. Phys. **34**, 427 (1984)

[8] H. E. Roman, J. Stat. Phys. **58**, 375 (1990)

[9] For more information see e.g. D. Stauffer, *Introduction to Percolation Theory*, Taylor and Francis, 1985

[10] L. Heupel, J. Stat. Phys. **42**, 541 (1986)

[11] O. Paetzold, J. Stat. Phys. **61**,491 (1990)

[12] H. P. Wittmann, K. Kremer, preprint. The update rate for this algorithm at a density of fifty per cent of the polymers on three-dimensional lattices is 2.2 updates per microsecond on a CRAY Y-MP.

# Ising Machine, m-TIS

*N. Ito[1], M. Taiji[1], M. Suzuki[1], R. Ishibashi[2], K. Kobayashi[2], N. Tsuruoka[2], and K. Mitsubo[3]*

[1]Department of Physics, Faculty of Science, University of Tokyo, Hongo, Bunkyo-ku, Tokyo 113, Japan
[2]Faculty of Science and Engineering, Tokyo Denki University, Hatoyama-cho, Hiki-gun, Saitama 350-03, Japan
[3]Department of Physics, Faculty of Science, Tokyo Institute of Technology, Ookayama, Meguro-ku, Tokyo 152, Japan

**Abstract** We have designed and constructed a special processor for Ising Monte Carlo simulation. The characteristics of this m-TIS are presented in this paper.

The study of many–body systems is interesting and one of the most important problems in modern physics. Monte Carlo simulation is one promising method of investigating the equilibrium state of many–body systems, although there are still many systems for which an effective Monte Carlo algorithm is not yet known. For most classical systems, the Monte Carlo method works effectively. If we have enough simulation power, we can estimate the physical quantities of the relevant system with good precision.

To get enough computational power, some people have been trying to make special computer systems[1]. We have also made a special computer system for Ising Monte Carlo simulations. Our system consists of a host computer and a simulation engine. The present host computer is a personal computer with 10MHz 80286 CPU. What we have designed and constructed is a simulation engine part, which is named the m-TIS, (mega-) *T*okyo University *I*sing *S*pin machine.

The m-TIS consists of local information registers (LIR), a random number generator (RNG), a spin flipper, a magnetization counter and a clock generator. When the m-TIS gets the start command from the host computer, the clock generator generates 16 clock pulses at 10MHz. The LIR keeps only some of the whole configurations which are necessary for updating 16 spins. It prepares and sends the local configurations necessary for updating one spin with the spin flipper synchronously to the system clock. Then it receives the next spin state from the spin flipper and stores it.

The spin flipper has the RAM table of transition probabilities which is initialized by the host computer. The local configuration is mapped to the transition probability and the transition probability is compared with the random number from the RNG to determine the next spin state.

The m-TIS has a magnetization counter, which counts the number of up spins without reducing the simulation speed. More details of the m-TIS are given in Ref.[1] and some results obtained by this system are given in Refs. [2] and [3].

The m-TIS is not a stand-alone-type simulator. It does not keep all configurations in itself. Spins, bonds or other configurations are all stored in the memory of the host computer. At each operation, necessary configuration data are transferred from the host computer. The start command is sent to the m-TIS and the updates of 16 spins start. After the updates, the next configurations of 16 spins are read by the host computer from the m-TIS and stored in the memory of the host computer.

There are several good points in this architecture arising from the separation of configuration memory from the special processor. The structure of the machine becomes simpler because the address generator and configuration memory are not necessary. The maximum system size is determined by the host computer, not by the m-TIS and therefore if we want to simulate a very large system we can do it by preparing a large memory in the host computer.

There are drawbacks in this architecture. The simulation speed is limited by the data transfer speed between the host and m-TIS. The clock cycle of the m-TIS is 10MHz, which means that it can simulate 10M spins per second if the data transfer speed is faster than the simulation speed of the m-TIS, more explicitly, if the necessary words and result word can be transferred within the simulation time of 16 spins, that is, $1.6\mu$sec. The necessary words are 5–7 for the ferromagnetic Ising model on a simple cubic lattice with nearest neighbor coupling. The variation 5–7 comes from the boundary conditions.

The total performance of the present system is, however, about 2M spins a second. It depends on the situation whether this speed is fast enough or too slow. If one wants to make the simulation faster, it is possible to construct the data transfer unit so that the data are transferred fast enough and to make the simulation at the maximum speed of the m-TIS.

We are now considering powering up the m-TIS so that it can treat more physically interesting systems. The address width for the transition probability table RAM of the present m-TIS is 13 bit. Therefore it can simulate a system whose local Hamiltonian contains less than or equal to 13 bits of information (one spin corresponds to one bit). We will increase it. The present physical quantity counter is only for the magnetization. The measurements of other quantities, for example the energy, are also expected. The number of such counters for physical quantities will be increased. The number of registers will be also increased. The random number generator will be improved. The details will be reported in the near future.

**Acknowledgments**

The present authors thank Prof. S. Katsura for supporting our studies. One of them ( N. I.) is grateful for financial support of the Fellowships of the Japan Society for the Promotion of Science for Junior Japanese Scientists. He also thanks the Inoue Foundation for Science for financial support to attend this workshop. This work is supported by the Research Fund of the Ministry of Education, Science and Culture.

# References

[1] N. Ito, M. Taiji and M. Suzuki, Rev. Sci. Instrum. , **59** (1988) 2483. See the references therein for other special purpose systems for theoretical physics.

[2] N. Ito, M. Taiji and M. Suzuki, J. Phys. Soc. Jpn. **56** (1987) 4218.

[3] N. Ito, M. Taiji and M. Suzuki, J. de Physique **49**(1988) 1397.

# Monte Carlo Analysis of Finite-Size Effects in the Three-Dimensional Three-State Potts Model

O.F. de Alcantara Bonfim

Department of Physics, University of Florida, Gainesville, FL 32611, USA

**Abstract.** A Monte Carlo study of the three-state Potts model in three dimensions is carried out. By using a histogram method recently proposed by Ferrenberg and Swendsen [Phys. Rev. Lett. **61**, 2635 (1988)], the finite-size dependence for the maximum of the specific heat is found to scale with the volume of the system, indicating that the phase transition is of first order. The value of the latent heat per spin and the correlation length at the transition are estimated.

## 1. Introduction

The knowledge of finite size dependence of the various thermodynamic quantities in the neighborhood of a phase transition, provides a very useful way to compute, using numerical techniques and appropriate extrapolation, the properties of infinite systems. In this paper we investigate the finite-size effects at a temperature-driven phase transition in the three-state Potts model in three dimensions. Numerical simulations here performed show that the maximum of the specific heat scales with the volume of the system in the thermodynamic limit indicating that the phase transition of the model is first order. We use a finite-size scaling form for the behavior of the maximum of the specific heat valid for all volumes to estimate the latent heat and correlation length at the transition point.

The model investigated here is described by the following Hamiltonian:

$$H = -J \sum_{<i,j>} \delta(\sigma_i, \sigma_j) - H \sum_i \delta(\sigma_i, 1) \qquad (1)$$

where $<i,j>$ indicates the sum over all pairs of nearest neighbors interactions. $\sigma_i = 1, 2, \ldots q$, specifies one of the q states of a spin at site i. The symbol $\delta(\sigma_i, \sigma_j)$ is the Kronecker $\delta$-function, $J > 0$ is the interaction energy between two spins and H is the applied magnetic field. This model has been extensively studied by differents techniques. Earlier studies were done by high-temperature series analysis [1-4], mean field theories [5], renormalization group techniques [6-8] and Monte Carlo simulations [9-11]. Recent calculations have been done mostly by numerical simulation using multilattice microcanonical simulation [12] and Monte Carlo techniques [13]. An extensive review of the static properties of the q-state Potts model in all dimensions is found in ref. [14].

## 2. The Histogram Method

By following Ferrenberg and Swendsen [15] let us rewrite the Hamiltonian (1) in the form

$$-\beta H = K \sum_{<i,j>} \delta(\sigma_i, \sigma_j) + h \sum_i \delta(\sigma_i, 1) \equiv KE + hM \qquad (2)$$

where $K$ is a dimensionless coupling constant in which we have absorbed the usual factor of $1/K_BT$, and $h = H/K_BT$. The partition function can now be written as

$$Z(K,h) = \sum_{E,M} N(E,M) exp(KE+hM) \qquad (3)$$

where N(E,M) is the number of states for the system with energy E and magnetization M. The probability distribution of E and M for a given temperature and field is given by

$$P_{(K,h)}(E,M) = N(E,M)\exp(KE+hM)/Z(K,h). \qquad (4)$$

So that the average of any thermodynamic quantity can be evaluated as

$$<A(E,M)>_{(K,h)} = \sum_{(E,M)} A(E,M) P_{(K,h)}(E,M). \qquad (5)$$

The probability distribution can be found numerically, by using Monte Carlo simulation to generate the histogram of values of E and M. The histogram properly normalized is an estimate of P(K, h) for a point (K, h) in the parameter space. The great advantage of the method is that the estimated probability distribution for a given point (K, h) can be used to generate the probability distribution for a different point (K', h') in the parameter space, that is

$$P_{(K',h')}(E,M) = \frac{P_{(K,h)}(E,M)\exp\left[\left(K'-K\right)E + \left(h'-h\right)M\right]}{\sum_{E,M} P_{(K,h)}(E,M)\exp\left[(K'-K)E + (h'-h)M\right]}. \qquad (6)$$

One should emphasize the the above relation is exact and any errors are due to the numerical determination of the number of states N(E). The denominator of eq. (6) is in fact an approximation for the partition function. Since we are interested in quantities that involve the various moments of the energy distribution, we shall restrict hereafter the phase space to one dimension involving only the energy E.

## 3. Finite Size Scaling

Before we report our numerical results let us first review the main results of the theory of finite-size scaling at temperature–driven first order phase transition, proposed by Challa et al. [16]. The starting point is the Landau theory of thermodynamic fluctuations [17]. In this approach the probability distribution P(E), of finding the system in a single phase with internal energy E, is a Gaussian centered about the infinite-lattice energy $E_0$, namely

$$P(E) = \frac{A}{\sqrt{C}} exp\left(-\frac{L^d(E-E_0)^2}{2K_BT^2C}\right), \qquad (7)$$

where A is a normalization constant and C is the specific heat at temperature T. In the thermodynamic limit the probability distribution P(E) is a $\delta$–function centered at $E_0$. Because of phase coexistence at a first order phase transition one can assume that the probability distribution P(E) is a superposition of Gaussians centered at the energies of each phase. Let

$E_+(E_-)$ be the internal energy at the transition in the high(low) – temperature phase. If the temperature is shifted by a small amount $\Delta T = T - T_c$ the probability distribution can be written as

$$P(E) = \frac{a_+}{\sqrt{C_+}} exp\left[-\frac{[E - (E_+ + C_+\Delta T)]^2 L^d}{2K_B T^2 C_+}\right]$$
$$+ \frac{a_-}{\sqrt{C_-}} exp\left[-\frac{[E - (E_- + C_-\Delta T)]^2 L^d}{2K_B T^2 C_-}\right] \quad (8)$$

where $C_+$ and $C_-$ are the discontinuity of the specific heat at the critical temperature, that is

$$C_\pm = \lim_{T \to T_c^\pm} C(T). \quad (9)$$

The weights $a_+$ and $a_-$ are functions of the free-energy difference $\Delta F = F_+ - F_-$ of the two phases:

$$a_\pm = \sqrt{C_\pm} exp(\pm x)$$
$$\text{with } x = -\frac{\Delta F}{2K_B T} L^d \quad \text{and} \quad \Delta F = -(E_+ - E_-)\Delta T/T_c. \quad (10)$$

In the special case when the transition occurs from a disordered state to a q-fold degenerate ordered state (as in the q-state Potts model), the expression for the probability distribution is identical to Eq. (8) except for $a_- = q\sqrt{(C_-)}exp(-x)$. From Eq. (8) it is straightforward to compute the moments of the energy distribution. In particular the specific heat is given by

$$C_L(T) = \frac{a_+ C_+ + a_- C_-}{a_+ + a_-} + \frac{a_+ a_- [(E_+ - E_-) + (C_+ - C_-)\Delta T]^2}{K_B T^2 (a_+ + a_-)^2} L^d. \quad (11)$$

From the above expression one finds that the maximum of the specific heat occurs at

$$\frac{T_c(L) - T_c}{T_c} = \frac{K_B T_c \ln\left[q\left(\sqrt{C_-/C_+}\right)\right]}{E_+ - E_-} L^{-d} \quad (12)$$

where $T_c(L)$ and $T_c$ are the transition temperatures of the finite and infinite systems, respectively. The maximum of $C_L$ behaves as

$$C_L^{max} \simeq \frac{(E_+ - E_-)^2}{4K_B T_c^2} L^d + \frac{C_+ + C_-}{2}. \quad (13)$$

From the above results one can see that the effect of finite size is to shift the temperature by an amount given by Eq. (12) and that it approaches the infinite temperature value as $L^{-d}$. The specific heat, on the other hand, diverges as $L^d$ with a slope proportional to the square of the latent heat.

## 4. Numerical Results

We have performed Monte Carlo simulations in the three dimensional three-state Potts model at the infinite-lattice transition temperature $T_c = 1.817 \pm 0.001$ and zero magnetic field, for lattices sizes between L= 3 and L= 15. As initial condition we have taken a configuration where all spins are in the same state $\sigma_i = 1$ and let the system evolve towards equilibrium under Glauber dynamics. The spins are flipped with the transition probability $p = exp[-\Delta E/K_B T_c]/[1+exp(-\Delta E/K_B T_c)]$ where $\Delta E$ denotes the change in the energy for a spin flip. The histogram of values of E was accumulated after discarding an appropriate number of Monte Carlo steps (MCS). Typical observation times were $2.5 \times 10^5 - 10^6$ MCS/spin, with the first $10^4$ MCS/spin discarded to allow for thermalization.

The thermodynamic quantities calculated in the present simulation are, the specific heat per spin, calculated from the energy fluctuation:

$$\frac{C}{K_B} = \beta^2 \left[ <E^2> - <E>^2 \right] /N \qquad (14)$$

and the reduced fourth order cumulant $V_L$, defined by

$$V_L = 1 - \frac{<E^4>}{3<E^2>^2}. \qquad (15)$$

For a first order phase transition $V_L$ takes on values 2/3 for low and high temperatures tending towards a non-trivial value of [16]:

$$V_L = \tfrac{2}{3} - \tfrac{1}{3} \left[ \frac{E_+^2 - E_-^2}{E_+^2 + E_-^2} \right]^2 \qquad (16)$$
$$L \to \infty$$

at $T_c$, where $E_+$ and $E_-$ are the discontinuity of the internal energy at the transition point. Using the estimates of Wilson and Vause [12] for the internal energy discontinuities $E_+ = -1.7831(8)$ and $E_- = -1.5862(6)$ we obtain for the cumulant at $T_c$ the value $V_L = 0.663$.

In Fig. 1 we show the distribution of the internal energy per spin for three different temperatures after about $10^6$ MCS through a $15^3$ lattice. The results show a double-peak structure in a small temperature range around $T = 1.817$ which is an indication of a first-order phase transition. Another indication that the transition is of first order is given by the temperature dependence of the fourth order cumulant $V_L$ as shown in Fig. 2. The value of $V_L$ at $T_c$ as the lattice size increases approaches the infinite value of $V_L = 0.663$ as obtained from Eq. 16. In Fig. 3 the maximum of the specific heat is plotted against the volume of the system. Each point shown in Fig. 3 has been calculated from the average of ten independent runs each measured using $2.4 \times 10^5$ MCS/spin. The error bars estimated from the dispersion of $C_L^{max}$ obtained by ten independent runs are smaller than the size of the open circles. The onset of the asymptotic behavior, that is, the maximum of the specific heat scaling with the volume, as predicted by Eq. 13, is already seen for large lattices size. Since the fully asymptotic behavior has not been reached for the lattices used in our simulation we can not use the data presented in Fig. 3 together with Eq.(13) directly to extract information about the latent heat and specific heat singularities at the transition point. In order to do that we must include correction to scaling terms into Eq.(13). We find by using a non-linear square fitting that the data is best fitted if a L-dependent exponential term is also added to the terms present in Eq.(13), as it was done for the two-dimensional Potts model [15]. In Fig. 4 is shown the maximum of the specific heat against L along with the non-linear best fit using

$$C_L^{max} = \alpha L^3 + \beta + \gamma exp(-L/l) \qquad (17)$$

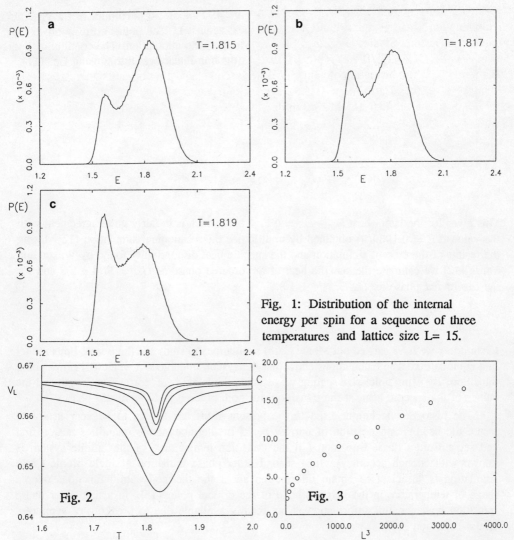

Fig. 1: Distribution of the internal energy per spin for a sequence of three temperatures and lattice size L= 15.

Fig. 2: Temperature dependence of the fourth order cumulant for the energy for lattices sizes L= 7, 9, 13 and 15. The lowermost curve corresponds to L= 7 and the uppermost to L= 15.

Fig. 3: Dependence of the specific heat maximum with $L^3$ for lattices sizes between L= 3 and L= 15.

where $\alpha$ is proportional to the square of the latent heat, as seen from Eq.(13) and $l$ is a measure of the correlation length. The best-fitted values for the above parameters are

$\alpha = 0.0023 \pm 0.0001$
$\beta = 12.9 \pm 2.0$
$\gamma = -13.8 \pm 1.9$
$l = 12.8 \pm 2.7$

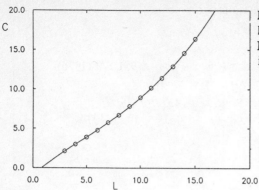

Fig. 4: Plot of the maximum of the specific heat against L. The points correspond to the Monte Carlo simulation. The continuous line is the non-linear best fitting using Eq. (17).

This gives for the latent heat $E_+ - E_- = 0.17 \pm 0.04$ which is in fairly good agreement with the value $\Delta E = 0.1969(9)$ obtained by multilattice microcanonical simulation [12]. Using the results of the present simulation and the specific heat discontinuity found by Wilson and Vause [12] we estimate the specific heat of the ordered phase as $C_- = 17.4 \pm 2.3$ and for the disordered phase as $C_+ = 8.4 \pm 1.8$.

## 5. Summary

In summary we have carried out Monte Carlo simulations to study the three-state Potts model in a cubic lattice. The calculations show that the system undergoes a weak first order phase transition. By using finite-size scaling we estimate the latent heat, correlation length and the values of the specific heat, at the transition, for both ordered and disordered phases.

The histogram technique used in the present work has shown to be very useful to locate the heights and position of narrow peaks of thermodynamic quantities near a first and second order phase transition ( if the transition temperature of the infinite system is known with enough accuracy). In a second order phase transition a single Monte Carlo simulation is sufficient to obtain information about the thermodynamic functions over a range of temperature in the neighborhood of the critical point [15]. In a first order phase transition however, this information is not readily available due to breaking of symmetry at the transition point.

## Acknowledgments

The author thanks Prof. Pradeep Kumar for useful suggestions and critical reading of the manuscript. This work was supported by DARPA under contract # MDA 972-88-J-1006.

## References

1. R. V. Ditzian and J. Oitmaa, J. Phys. A **7**, L61 (1974).
2. J. P. Straley, J. Phys. A**7**, 2173 (1974).
3. I. G. Enting, J. Phys. A **7**, 1617 (1974).
4. I. G. Enting and C. Domb J. Phys. A **8**, 1228 (1974).

5. S. Alexander, Solid State Commun **14**, 1069 (1974).
6. E. Golner, Phys. Rev. B **8**, 3419 (1973).
7. J. Rudnick, J. Phys. A8, 1125 (1975).
8. T. W. Burkhardt, H. J. F. Knops, and M. den Nijs, J. Phys. A **9**, L179 (1976).
9. H. J. Herrmann, Z. Physik B **35**, 171 (1979).
10. H. W. J. Blote and R.H. Swendsen, Phys. Rev. Lett. **43**, 799 (1979).
11. S. J. Knak Jensen and O. G. Mouritsen, Phys. Rev. Lett. **43**, 1736 (1979).
12. W. G. Wilson and C. A. Vause, Phys. Rev. B36, 587 (1987).
13. M. Fukugita and M. Okawa, Phy. Rev. Lett. **63**, 13 (1989). R. V. Gavai, F. Karsch, and B. Petersson, Nucl. Phys. **B322[FS3]**, 738 (1989).
14. F. Y. Wu, Rev. Mod. Phys. **54**, 1 (1982).
15. A. M. Ferrenberg and R. H. Swendsen, Phys. Rev. Lett. **61**, 2635 (1989). Earlier works using the histogram method are: Z. W. Salsburg, W. Fickett, and W. W. Wood, J. Chem. Phys. **30**, 65 (1959); I. R. McDonald and K. Singer, Discuss. Faraday Soc. **43**, 40 (1967). Further discussion of the present method is found in A. M. Ferrenberg and R. H. Swendsen Phys. Rev. Lett. **63**, 1195 (1989). See also the latter reference for a more complete list of works using the histogram method.
16. M. S. S. Challa, D. P. Landau, and K. Binder, Phys. Rev. B **34**, 1841, (1986)
17. L. D. Landau and E. M. Lifshitz, *Course of Theoretical Physics*, Vol. 5 of *Statistical Physics* (Pergamon, New York, 1985).

# Simulation Studies of Oxygen Ordering in $YBa_2Cu_3O_{7-\delta}$ and Related Systems

Zhi-Xiong Cai* and S.D. Mahanti

Department of Physics and Astronomy and
Center for Fundamental Materials Research,
Michigan State University, East Lansing, MI 48824, USA
*Present address: Department of Applied Science, Building 480,
Brookhaven National Laboratory, Upton, NY 11973, USA

**Abstract.** Monte Carlo simulation with constant oxygen concentration of a lattice gas model with anisotropic interaction is used to explore the effect of thermal quenching on oxygen ordering in the basal plane of $YBa_2Cu_3O_{7-\delta}$ and related systems. The effect of substituting $Ga^{3+}$ or $Al^{3+}$ ions at $Cu(1)$ chain sites on the oxygen ordering is also discussed within the framework of this model.

## 1 Introduction

It is now well recognised that the superconductivity in recently discovered high-$T_c$ oxide superconductors originates in the $CuO_2$ planes [1]. In the $YBa_2Cu_3O_{7-\delta}$ systems (1:2:3 compounds), in addition to the $CuO_2$ planes, there are oxygen deficient $CuO_{1-\delta}$ ($0 < \delta < 1$) planes referred to as the basal planes. Whereas in the $CuO_2$ planes oxygen atoms form a square network, in the basal planes they form chains along the so-called b-axis. In the 1:2:3 compounds, holes in both the $CuO_2$ planes and in the CuO chains contribute to conductivity and, in fact, in the absence of CuO chains the material is not a superconductor. These observations have led to extensive study of the thermodynamics of oxygen ordering in the basal planes for study of the entire range of $\delta$ values mentioned above [2,3,4]. In addition, some attempts have been made to relate the basal plane oxygen ordering to the hole concentration in the $CuO_2$ plane [5]. Furthermore, to elucidate the role of chain and plane Cu in superconductivity, several experimental and theoretical studies have been performed in systems where Cu is replaced by either trivalent (Ga, Al) or divalent (Zn) ions [6].

In this paper we discuss the thermodynamics of oxygen ordering and the associated high-T tetragonal to low-T orthorhombic structural transition in the 1:2:3 and related systems. In particular we discuss the success of anisotropic lattice gas models in describing the experimental results.

## 2 Anisotropic Short Range Lattice Gas Model

The basic assumption underlying the lattice gas model is that the complex Hamiltonian involving oxygen, copper and other atoms can be replaced by an effective short range anisotropic lattice gas (or Ising) Hamiltonian $H_{LG}$ given by [7]

$$H_{LG} = V_1 \sum_{nn} n_i n_j + V_2 \sum_{nnnb} n_i n_j + V_3 \sum_{nnna} n_i n_j + (\epsilon - \mu) \sum_i n_i \qquad (1)$$

where

$$\epsilon - \mu = E_O + E_d/2 + \alpha x - k_B T \left( \frac{P(1 - exp(-\frac{T_\nu}{T}))}{P_0 T^{7/2}} \right)^{1/2}, \qquad (2)$$

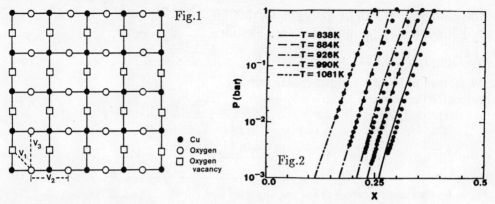

Fig.1. The ground state structure of the CuO plane of the YBa$_2$Cu$_3$O$_7$ system.

Fig.2. Pressure dependence of oxygen concentration isotherm obtained by Monte Carlo simulation [7] compared with the experimental data by Salomons et al. [8], the dots are experimental data.

where $E_O$ is the binding energy of the oxygen atom in the basal plane, $E_d$ is the dissociation energy of an oxygen molecule and $x$ is the average oxygen site occupancy; $n_i(=0,1)$ measures the occupancy of the $i$th lattice site by an oxygen atom. The effect of Cu and other atoms has been replaced by an effective interaction between oxygen atoms $V_i$ shown schmatically in Fig. 1. The nearest neighbor (nn) interaction $V_1 > 0$, due to oxygen-oxygen repulsion. We assume $V_2 < 0$ and $V_3 > 0$ [7], these being physically reasonable for these systems due to Cu-O covalence bonding. Monte Carlo simulation of this model was shown to give an accurate picture of the oxygen pressure $(P)$ and temperature $T$ dependence of the basal plane ordering [7].

In Fig. 2, we show the $P$ vs $x$ for different $T$ obtained with the choice $V_1 = 0.25 eV$, $V_2/V_1 = -0.8$ and $V_3/V_1 = 0.4$ and using a constant $P$ (open system) Monte Carlo simulation. These are compared with the experimental results of Salomons et al. [7,8]. This set of parameters also reproduces the $P$ dependence of the oxygen ordering temperature (equal to the tetragonal-orthorhombic structural transition temperature).

## 3  Thermal Quenching and Oxygen Ordering

Extensive Monte Carlo simulations with constant $x$ (using Kawasaki dynamics) [9] have been performed to understand the nature of oxygen ordering for $x = 0.25, 0.35$ and to explore the relationship between the quenching rate and the low-$T$ oxygen structure. For $x = 0.25$ the simulations give a "double cell" orthorhombic structure with half the CuO chains intact and half empty.

The annealed and quenched results for $x = 0.35$ are shown in Fig. 3. Since this oxygen concentration is not appropriate for either the structure shown in Fig. 1 (which requires $x = 0.5$) or the "double cell" structure, new features arise. From Fig. 3(a) we find that the low-temperature structure for the annealed system is a highly degenerate one. The long CuO chains are intercalated into each other. This is quite different from the configuration found in constant chemical potential simulations [10] which show phase separation of the two ordered structures. The difference is due to the slow dynamics of oxygen diffusion below 400K [4]. The equilibruim state is harder to reach by the Kawasaki dynamics. In

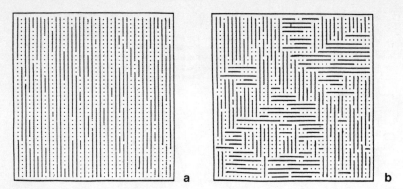

Fig.3. Oxygen configuration of (a) the annealed and (b) the quenched $YBa_2Cu_3O_{6.7}$ system.

the rapidly quenched systems, MC simulation gives short chains consisting on the average of 5-10 oxygen atoms oriented along both $a$ and $b$ axes, the resulting configuration showing tetragonal (T) symmetry as shown in Fig. 3(b).

## 4 Effect of Impurities on the Oxygen Ordering

Recently, several authors [6] have explored the effects of replacing Cu ions by other cations to elucidate the role of chain and plane Cu ions on the observed superconducting and structural properties. It is known that the Cu ions belonging to the chain sites (Cu(1) sites) can be preferentially substituted by nonmagnetic elements like Ga and Al. X-ray and neutron diffraction studies indicate that the structure of $YBa_2Cu_{3-x}(Ga,Al)_xO_7$ becomes tetragonal for $x > 0.18$ at room temperature. Unlike the tetragonal phase of rapidly quenched $YBa_2Cu_3O_{7-\delta}$, the tetragonal phase of this compound is superconducting with transition temperature $\sim$80K. We use an extended lattice gas model based on the assumptions that: (a) the binding energy of the oxygen is increased near the impurity site and (b) the oxygen-oxygen repulsion is increased near the impurity site either due to the increase of the negative charge of oxygen ions or due to the decrease in local screening because of the localization of the holes. This modified lattice gas model indicates the formation of local "cross-links" of oxygen ions near the site of Ga or Al impurities as shown in Fig. 4. Fig. 5 gives the oxygen configuration of the $YBa_2Cu_{3-x}(Ga,Al)_xO_7$ system for $x = 0.12$ obtained from Monte Carlo simulation at 300K. It shows that the cross-links destroy the

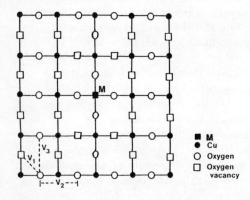

Fig.4. The structure of the basal plane of the $YBa_2Cu_{3-x}M_xO_7$ system with a single impurity M.

Fig.5. Oxygen configuration of the $YBa_2Cu_{3-x}M_xO_7$ system for $x = 0.12$ obtained from Monte Carlo simulation at 300K.

Cu-O chains along the b-axis and form short Cu-O chains along the a-axis. For large concentrations of Ga or Al, the system becomes tetragonal. This model gives an excellent account of the experimentally observed $x$-dependence of the oxygen ordering at 300K in the above systems [9].

## 5 Summary

In summary, states of partial oxygen order in $YBa_2Cu_3O_{7-\delta}$ and $YBa_2Cu_{3-x}M_xO_7$ systems have been obtained using a short range anisotropic lattice gas model. This model adequately describes the tetragonal to orthorhombic structural transition and can explain many experimental facts. Additional theoretical work is necessary to understand the complete phase diagram for this interesting model, especially at low temperature.

## Reference

1. P.W.Anderson, *Kathmandu Summer School Lectures*, 1989.
2. J.D.Jorgensen, M.A.Beno, D.G.Hinks, L.Sorderholm, K.J.Volin, R.L.Hitterman, J.D.Grace, I.K.Schuller, C.U.Segre, K.Zhang, and M.S.Kleefisch, Phys. Rev.B**36**, 3608 (1987).
3. R.J.Cava, C.H.Chen, E.A.Rietman, S.M.Zahurak, and D.J.Werder, Phys. Rev. B**36**, 5729 (1987).
4. C.H.Chen, D.J.Werder, L.F.Schneemeyer, P.K.Gallager and J.V. Wazczak, Phys. Rev. B**38**, 2888 (1988).
5. J.Zaanen, A.J.Paxton, O.J.Jepsen, and O.K.Anderson, Phys. Rev. Lett.**60**, 2685 (1988).
6. Gang Xiao, M.Z.Cieplak, D.Musser, A.Gavrin, F.H.Seitz, and C.L. Chien, Phys. Rev. Lett.**60**, 1446 (1988).
7. Zhi-Xiong Cai and S.D.Mahanti, Sol.St.Comm. **67**,287 (1988).
8. E.Salomons, N.Koeman, R.Brouwer, D.G. de Groot and R.Griessen, Sol.St.Comm. **64**, 1141 (1987).
9. Zhi-Xiong Cai and S.D.Mahanti, Phys.Rev.B **40**, 6558 (1988).
10. D.de Fontaine, M.E.Mann, and G.Ceder, Phys.Rev.Lett.**63**, 1300 (1989).

# Index of Contributors

Banavar, J.R. 65
Binder, K. 4, 172
Bouzida, D. 193

Cai, Zhi-Xiong 210
Carmesin, H.-O. 183

de Alcantara Bonfim, O.F. 203
Dünweg, B. 85

Ferrenberg, A.M. 30
Fraser, D.P. 99

Gerling, R.W. 43
Grest, G.S. 85

Halley, J.W. 187
Hjort Ipsen, J. 99

Ishibashi, R. 201
Ito, N. 16, 201

Johnson, B. 187
Joynt, R. 116
Jørgensen, K. 99

Klein, W. 50
Kobayashi, K. 201
Koplik, J. 65
Kremer, K. 85
Kumar, S. 193

Landau, D.P. 1, 183

Mahanti, S.D. 210
Manousakis, E. 123
McRae, W.B. 156
Mitsubo, K. 201
Mon, K.K. 1
Moreo, A. 138
Mouritsen, O.G. 99

Novotny, M.A. 177

Paetzold, O. 197

Reger, J.D. 172

Schüttler, H.-B. 1
Scheucher, M. 172
Suzuki, M. 16, 201
Swendsen, R.H. 193

Taiji, M. 201
Tsuruoka, N. 201

Vallés, J.L. 187

White, S.R. 145
Willemsen, J.F. 65

Young, A.P. 172

Zuckermann, M.J. 99

Printing: COLOR-DRUCK DORFI GmbH, Berlin
Binding: Buchbinderei Lüderitz & Bauer, Berlin

9062003
6400
250